Green Organic Chemistry and Its Interdisciplinary Applications

Green Organic Chemistry and Its Interdisciplinary Applications

Vera M. Kolb
University of Wisconsin-Parkside
Kenosha, USA

CRC Press
Taylor & Francis Group
Boca Raton London New York

CRC Press is an imprint of the
Taylor & Francis Group, an **informa** business

CRC Press
Taylor & Francis Group
6000 Broken Sound Parkway NW, Suite 300
Boca Raton, FL 33487-2742

First issued in paperback 2020

© 2016 by Taylor & Francis Group, LLC
CRC Press is an imprint of Taylor & Francis Group, an Informa business

No claim to original U.S. Government works

ISBN 13: 978-0-367-57482-6 (pbk)
ISBN 13: 978-1-4987-0207-2 (hbk)

Library of Congress Cataloging-in-Publication Data

Names: Kolb, Vera M.
Title: Green organic chemistry and its interdisciplinary applications / Vera M. Kolb.
Description: Boca Raton : Taylor & Francis, 2016. | "A CRC title." | Includes bibliographical references and index.
Identifiers: LCCN 2016008937 | ISBN 9781498702072 (hardcover : alk. paper)
Subjects: LCSH: Green chemistry--Textbooks. | Chemistry, Organic--Textbooks.
Classification: LCC TP155.2.E58 K65 2016 | DDC 660--dc23
LC record available at http://lccn.loc.gov/2016008937

Visit the Taylor & Francis Web site at
http://www.taylorandfrancis.com

and the CRC Press Web site at
http://www.crcpress.com

Dedication

In loving memory of my parents, Dobrila and Martin Kolb.

Contents

Preface

Green organic chemistry is developing at a fast pace. It is increasingly being applied to other branches of chemistry, but also to the chemical and pharmaceutical industries, and to the field of engineering. It is also becoming increasingly present in the undergraduate chemistry curriculum. It is my hope that this book will be useful to students, instructors, and chemists in general in learning about green organic chemistry and its interdisciplinary applications.

This book reflects my enthusiasm for green chemistry, which I wish to transfer to students and other readers.

The book consists of 13 chapters. Each chapter has sections on learning goals, review problems, answers to the review problems, and references.

Chapter 1, "Introduction to Green Chemistry," covers history of the emergence and establishment of green chemistry, as well as its innovative aspects.

Chapter 2, "Twelve Principles of Green Chemistry," describes these principles in depth, typically on specific examples.

Chapter 3, "Innovative Aspects of Green Chemistry," gives various examples of innovations in green chemistry, and also describes different types of thinking that promote creativity and innovation.

Chapter 4, "Green Organic Reactions 'on Water,'" covers reactions of organic compounds that are not water soluble, but nonetheless react in the aqueous medium. Various examples of such reactions are given.

Chapter 5, "Green Organic Reactions in Superheated Water," shows additional possibilities that water has as a green solvent. This chapter provides foundation of fundamental properties of water, phase diagram of water, properties of superheated water, and it gives various examples of organic reactions in superheated water.

Chapter 6, "Green 'Solventless' Organic Reactions," covers another green way of doing organic experiments, namely without adding solvents. Solventless reactions that are based on the melting behavior of the reaction mixtures are discussed, and examples of such reactions are provided.

Chapter 7, "Green Organic Reactions in the Solid State," describes how solid-state reactions occur and provides various examples of such reactions, including photochemical reactions, microwave-assisted reactions, and rearrangements.

Chapter 8, "Applications of Green Chemistry Principles to Engineering: Introduction to Sustainability," first provides a background of green engineering and then describes all of its 12 principles in depth. Principles of green chemistry and green engineering are then compared. Sustainability as related to green engineering is covered, including the natural step, biomimicry, and cradle-to-cradle approaches. Descriptions of systems' and holistic thinking, which are two types of thinking that are useful in green engineering, are also provided.

Chapter 9, "Chemical Industry and Its Greening: An Overview of Metrics," first gives a background of chemical industry in general and then it focuses on polymer industry. Specific examples of greening of the chemical industry are given. This chapter also introduces green chemistry metrics.

Chapter 10, "Applications of Green Chemistry Principles in the Pharmaceutical Industry," first provides a background of the pharmaceutical industry and then covers difficulties that pharmaceutical industry often has in meeting the green chemistry goals. Specific examples, including those on the greening of the pharmaceutical processes, are also provided.

Chapter 11, "Applications of Green Chemistry Principles in Analytical Chemistry," first covers the application of the 12 principles of green chemistry to analytical chemistry and then GAC (green analytical chemistry) with its specialized green principles. Various examples of greening of analytical chemistry are also given.

Chapter 12, "Application of Green Chemistry Principles in Environmental Chemistry," provides a background of environmental chemistry and discusses relationship between the environmental chemistry and green chemistry. Pharmaceuticals are discussed as an example of emerging environmental pollutants. Finally, the developments of ecofriendly chemicals in response to the findings of the environmental chemistry are also described.

Chapter 13, "Greening Undergraduate Organic Laboratory Experiments," provides a review of the literature sources on green organic laboratory experiments and presents selected methods of greening organic laboratory experiments. These methods are discussed as the examples of specific experiments.

This project would not have been possible without the help of the CRC Press/ Taylor & Francis Group's excellent editorial and publishing staff, to whom thanks are expressed. Special thanks are due to Francesca McGowan and Hilary LaFoe, two outstanding acquisition editors.

Vera M. Kolb
University of Wisconsin-Parkside

Acknowledgments

Most sincere thanks are expressed to my colleague and friend, Professor Richard H. Judge, for his editorial help, comments on the content of the book, and suggestions for its improvement.

Author

Vera M. Kolb earned a chemical engineering degree and an MS in chemistry at Belgrade University, Serbia. She pursued her chemistry studies as a Fulbright scholar, and earned a PhD at Southern Illinois University at Carbondale. She is currently a professor of chemistry and the director of the Center for Environmental Studies at the University of Wisconsin, Parkside. She had two sabbatical leaves, one at the Salk Institute for Biological Studies and the other at Northwestern University. She has over 150 publications, including two books and three patents, in the fields of organic reactions and mechanisms, morphine-type opiates and their receptors, estrogens, nucleosides, teratogens, organic reactions in water and in the solid state, chemistry under prebiotic conditions, astrobiology, and chemical education. In 2002 she was inducted into the Southeastern Wisconsin Educators' Hall of Fame, and in 2013 she received the Phi Delta Kappa Outstanding Educator Award.

1 Introduction to Green Chemistry

The green hat is for creative thinking.

Edward de Bono (1985)

... green chemistry has also been described as the Hippocratic Oath for the Chemist ("First, do no harm") ...

Paul T. Anastas and John C. Warner (1998)

LEARNING OBJECTIVES

The learning objectives for this chapter are as follows.

Learning Goals	Section Numbers
History of green chemistry	1.1
Historical setting for the emergence of green chemistry	1.1.1
Emergence and establishment of green chemistry	1.1.2
Some common descriptions of green chemistry	1.2
Innovative aspects of green chemistry	1.3

1.1 A BRIEF HISTORY OF GREEN CHEMISTRY

The term "green chemistry" was introduced by Paul Anastas in 1991. In most historical accounts of green chemistry, this year is marked as the official beginning of green chemistry. However, the understanding of green chemistry would be incomplete unless we consider the historical setting for its emergence.

1.1.1 HISTORICAL SETTING FOR THE EMERGENCE OF GREEN CHEMISTRY

Rapid advances in chemistry in the last century brought numerous benefits to the society. New drugs were developed which eradicated or alleviated many diseases. Agriculture has become much more productive due to the new agricultural chemicals. Effective pest control became possible by the use of the chemically manufactured pesticides. Natural textiles and dyes were rapidly substituted with the less expensive and often better products that resulted from chemical synthesis. Introduction of plastics and other polymer-based products has revolutionized manufacturing of the automobile parts, fluid transport pipes, furniture, and various other household products, among many other uses. Medicine has benefited from the new materials, such as those used in artificial hips and various stents. Car paints, house paints, personal care products, such as shampoos, and flame retardants used in the furniture fabrics are among many examples of the benefits that chemistry has brought to the society.

Many of the beneficial chemicals such as pharmaceuticals, agricultural chemicals, and various plastics were obtained by organic syntheses that involve toxic chemicals. The latter could be toxic starting materials, reagents, catalysts, solvents, and by-products, which could end up as a toxic waste. Particularly, dangerous are toxic volatile organic substances, because they can escape to the environment faster than most solid materials or high boiling liquids. Toxic materials of all sorts were dumped into the rivers or other waters, under the assumption that they would be eventually diluted to the point that they would not be hazardous anymore. Likewise, the volatile toxic chemicals were often released to the atmosphere, again believing that dilution in the air would render them nonhazardous. A large amount of toxic chemical waste was buried in the land sites, directly or in the containers. Such practices were considered safe.

However, several major environmental disasters were caused by the release of toxic chemicals into the environment, thus challenging the presumed safety of the above practices. Selected examples of toxic chemicals and the associated environmental disasters are given in Table 1.1. The chemical names and structures for the chemicals from Table 1.1 are shown in Figure 1.1.

TABLE 1.1
Examples of Toxic Chemicals and Associated Environmental Disasters

Chemical (Common Name)	Use/Source	Environmental Disaster (a) Description (b) Location (c) References/Action/Status
DDT	Insecticide	(a) Toxic to many organisms (humans, wild life, such as birds) other than to the targeted pests
		(b) Widespread
		(c) Rachel Carlson's *Silent Spring*, 1962; DDT was banned in the United States in 1972.
Dioxin	Industrial by-product	(a) Site contamination with dioxin from industrial waste oil in the early 1970s
		(b) Times Beach, Missouri
		(c) Residents were evacuated in 1983–1985.
Dioxin	Industrial waste	(a) Site contamination by leak of buried toxic waste in 1978
		(b) Love Canal, near Niagara Falls, New York
		(c) Residents were relocated.
MIC	Reagent in synthesis of Carbaryl (an insecticide)	(a) Leak of more than 40 tons of MIC, a toxic gas, from a pesticide plant; at least 3000 people died and hundreds of thousands were injured; December 3, 1984
		(b) Bhopal, India
		(c) Carbaryl synthesis was redesigned to avoid preparation of MIC.

Note: See Figure 1.1 for the structures and chemical names.

DDT*
Chemical name: Dichlorodiphenyltrichloroethane

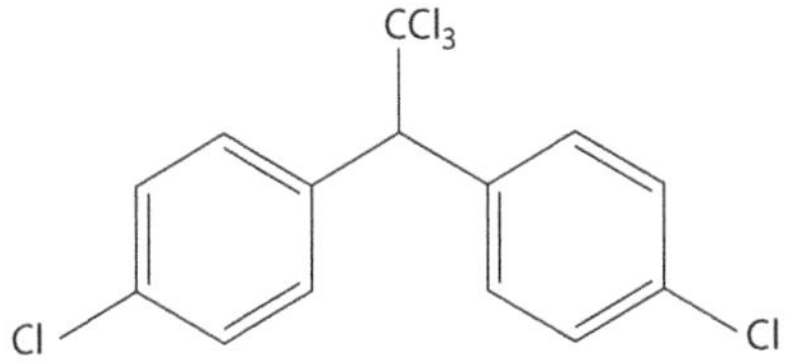

Dioxin*
Chemical name: 2,3,7,8-Tetrachlorodibenzo-*p*-dioxin

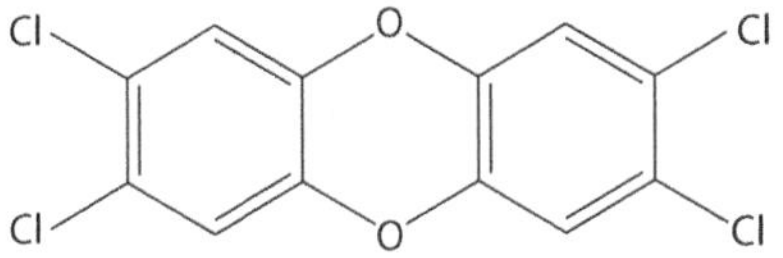

MIC
Chemical name: Methyl isocyanide

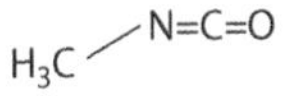

FIGURE 1.1 Chemical structures and names of chemicals from Table 1.1. (The principal toxicants, often mixed with closely related compounds, are labeled with *.)

The entries from Table 1.1 are briefly addressed. Dichlorodiphenyltrichloroethane (DDT), an effective insecticide, caused unintentional damage to other organisms, including humans and wild life, such as birds. In her 1962 book, *Silent Spring*, Rachel Carlson brought attention to this problem. Ten years later, DDT was banned in the United States. Table 1.1 includes information on two major environmental disasters, which were caused by contamination by dioxin, in the 1970s and 1980s. Two contaminated sites, Times Beach, Missouri, and Love Canal, near Niagara Falls, New York, had to be evacuated and the residents relocated. The Love Canal incident ultimately led to the passage of federal legislation governing hazardous waste sites. The last entry in Table 1.1 informs us about a terrible tragedy in Bhopal, India, in 1984, in which many thousands of people died or were injured. The chemical responsible for this tragedy was methyl isocyanide (MIC), a toxic gas. It was released accidentally.

In this period of history, which lasted roughly through the mid-1980s, the prevailing thinking was as follows: Chemicals are often toxic. Dangerous situations may arise unexpectedly when we work with such chemicals. What we need to do is to devise better and more comprehensive methods for protection against toxic chemicals.

However, as we have seen from the given examples of environmental disasters, what may seem like a good protection method at the time may fail for a variety of unforeseen reasons. Containers in which chemicals are buried may become corroded over a period of time and may start leaking the chemicals. The buried chemicals may surface during flooding. Accidental release of toxic chemicals may happen due to the human error, but also due to the natural causes, such as floods, fires, and earthquakes, to name a few. It seems that building more and better protection against toxic chemicals may be a very expensive proposition, which may not be feasible or effective in some cases, and it does not appear to guarantee a full-proof protection.

1.1.2 EMERGENCE AND ESTABLISHMENT OF GREEN CHEMISTRY

The late 1980s and the early 1990s was the historical period during which green chemistry emerged. As it is often the case, big problems in science may be solved by novel and innovative thinking, which liberates itself from the old beaten paths on how things were done in the past. Such new thinking often results in bold and groundbreaking solutions. This is how green chemistry was born. Paul Anastas proposed that chemistry should be done by using benign chemicals and processes. This proposed paradigm represented a shift from protection from toxic chemicals to the avoidance of toxic chemicals to begin with. The proposed focus was on using benign chemicals instead of the toxic one, on the development of benign chemical processes and on the redesign of the old chemical processes to make them more benign. The term "green chemistry" was introduced in 1991 by Paul Anastas to describe this new paradigm.

The Pollution Prevention Act of 1990 reflects the regulatory policy change from pollution control to pollution prevention as a more effective method. This act incorporates the green chemistry paradigm, which states that the focus should be on the benign chemicals/processes, as they prevent pollution.

Green chemistry premises were gradually accepted. Selected mileposts in the evolution of green chemistry are given.

In 1995, the Presidential Green Chemistry Challenge Award was created for scientific innovation in academia and industry that advance green chemistry. In 1997, the first PhD program in green chemistry was established (at the University of Massachusetts in Boston, Massachusetts). In the same year, the Green Chemistry Institute was founded. A solid guidance to the green chemistry paradigm was given in 1998 by Anastas and Warner who formulated and published the 12 principles of green chemistry. In 2000s, various green chemistry groups were formed, journals dedicated to the green chemistry were founded, and green chemistry conferences all over the world were taking place. Today, green chemistry is an established and rapidly evolving field.

1.2 SOME COMMON DESCRIPTIONS OF GREEN CHEMISTRY

In this section, the paradigm of green chemistry is built up, which was explained in Section 1.1.2. More complex descriptions are gradually introduced. These will serve as a necessary background for the next chapter, which will cover the 12 principles of green chemistry in detail. Some frequently used descriptions of green chemistry are presented below, some of which cover only its selected aspects.

From Anastas and Williamson (1996)

- Green chemistry eliminates hazards by redesigning the experiments, rather than keeping hazardous experiments, but preventing exposure.
- Green chemistry is often referred to as "environmentally benign chemistry" and "benign-by-design chemistry."
- Green chemistry is also referred to as an approach to chemistry, which eliminates or reduces risks to human health and environment.

TABLE 1.2

Concise and Simplified Version of 12 Principles of Green Chemistry

1. Waste prevention
2. Atom economy
3. Safer syntheses
4. Safer products
5. Safer auxiliaries (such as solvents)
6. Energy efficiency
7. Renewable feedstock
8. Derivative reduction
9. Catalysis
10. Degradability
11. Pollution prevention
12. Accident prevention

Source: Sanderson, K., It's not easy being green, *Nature*, 469, 18–20, 2011. With permission.

- Green chemistry is a new paradigm in chemistry in which new chemistry is invented and put into practice, with an emphasis on the new synthetic methods, reaction conditions, catalysts, analytical tools, and industrial processes, which are all nonhazardous and environmentally benign.
- "Green Chemistry is the use of chemistry techniques and methodologies that reduce or eliminate the use or generation of feedstocks, products, by-products, solvents, reagents, etc., that are hazardous to human health or the environment" (p. 2).

From Tundo and Aricò (2007)

- Green chemistry is defined by the International Union of Pure and Applied Chemistry as follows: "The invention, design, and application of chemical products and processes to reduce or eliminate the use and generation of hazardous substances."

Anastas and Warner (1998) formulated the 12 principles of green chemistry, which is covered in detail in Chapter 2 and throughout this book. A short version of these principles is introduced in Table 1.2.

1.3 INNOVATIVE ASPECTS OF GREEN CHEMISTRY

From Section 1.2, we have seen that the emergence of green chemistry was the result of innovation. Green chemistry changed the existing ways chemistry was done, as it turned from protection from chemical hazards to their elimination.

Green chemistry requires creative work to invent and put into practice chemical reactions and processes to realize its principal tenet of being a benign chemistry.

However, a large number of chemical reactions and processes that are not green are still in place. It would be impractical, costly, and in some cases impossible to replace them with the green processes on a short notice. One major impediment is that the required green processes may not exist and are yet to be discovered.

Still, green chemistry changes may be implemented to some aspects of the already existing chemical reactions and processes via "greening" them. This means that we can make some aspects benign or less hazardous. Although the overall process will still have some hazardous features, it will be less hazardous than before the greening.

Both the invention of the brand new green chemical reactions and processes and the greening of the already existing ones require innovative thinking. However, complex thinking is also required.

What do we mean by complex thinking in this context? We have briefly introduced the 12 principles of green chemistry. If we think of just one, and pursue this single principle path, this would be a case of the linear thinking. Such a linear path may not be sufficient: we may have to simultaneously consider all other principles and take them into account to produce the best green solution.

As one example, when we are focusing on the synthesis of a product that needs to be nontoxic to human health and safe to the environment (principle 3), we need to take into account the energy requirements for such a synthesis (principle 6), and be aware that any incomplete conversion of starting materials would create waste, which would go against principle 1. (See Table 1.2 for the principles.)

A thorough knowledge of the 12 principles of green chemistry is thus necessary not only at the level of the individual principles but also at the level of their often complex interactions.

REVIEW QUESTIONS

1.1 Give five examples of beneficial uses of chemistry other than those listed in the text.

1.2 Give the names of chemicals from the products below, which are easily available in your home, as found on the labels of the products. Then search these names on the Internet to find the chemical structures and the alternative chemical names. The products are as follows: (a) Preservatives listed on cereal boxes, usually under "food additive to preserve freshness"; active ingredients in the over-the-counter painkillers (b) Advil® and (c) Tylenol. This exercise will enforce the positive habit of correlating the chemical name to its structure.

1.3 Name some other dioxin exposure incidents other than those listed in Table 1.1. They can be easily found on the Internet by searching the keyword "polychlorinated dibenzodioxins," which is the category of chemicals to which dioxin belongs.

1.4 Examine the statements below to establish if they reflect the green chemistry approach. Answer "Yes" or "No," but justify your answer.

    a. A reaction vessel was redesigned to introduce a backup safety valve, which would shut off the accidental release of the toxic gases to the environment.

    b. A toxic product was found to be rapidly degraded in the environment, and was therefore considered safe for the burial at the land site.

c. A manufacturing process produced chemical waste, which was then used as a starting material in another process.

1.5 Group exercise: Prepare index cards with the 12 principles of green chemistry. Mix up the cards and randomly select three cards at any time. Try to draw a correlation between the principles. If only one principle is followed, would this compromise the other two?

ANSWERS TO REVIEW QUESTIONS

1.1 Answers will vary. Some examples are given here: (1) artificial sweeteners, which benefit people with diabetes; (2) batteries, which are used in cars, watches, and so on; (3) drugs used for veterinary purposes; (4) decaffeinated coffee, which is obtained from natural coffee by extraction of caffeine; (5) algaecides, which are used for control of algae in the swimming pools.

1.2 (a) Butylated hydroxytoluene; (b) Advil®: Ibuprofen; (c) Tylenol: Acetaminophen. The structures are shown in Figure 1.2.

1.3 Some examples are as follows: industrial accident in Seveso, Italy, in 1976; contamination of animal feed with dioxin in Belgium in 1999; and Irish pork crisis in 2008, due to the high levels of dioxin in pork.

1.4 a. No. This is a case of prevention of hazard, rather than its elimination.

   b. No. The degradation products could also be toxic. Even if not toxic by themselves, they could react with other compounds that are buried to give toxic material. Until this is established, the practice is not safe. Other reasons may exist.

   c. Yes. The waste that may be unavoidable for a specific process would be used for a beneficial purpose.

1.5 Answers will vary, depending on the choice of cards (principles).

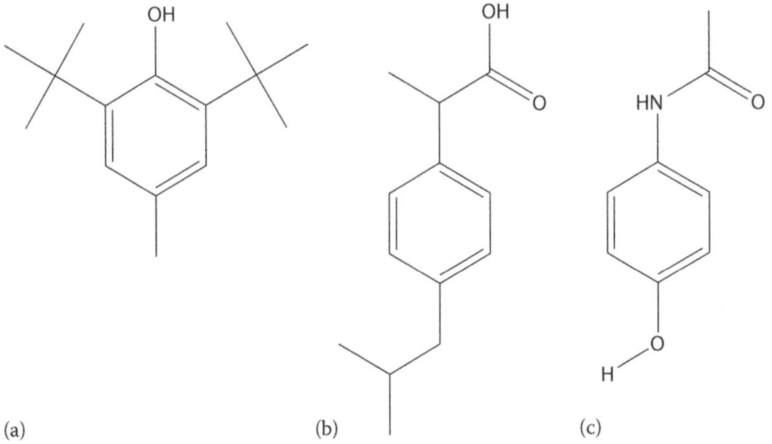

(a) (b) (c)

FIGURE 1.2 Structures of compounds for question 1.2: (a) Butylated hydroxytoluene; (b) Advil®: Ibuprofen; and (c) Tylenol: Acetaminophen.

REFERENCES

Anastas, P. T. and Warner, J. C. (1998). *Green Chemistry: Theory and Practice*, Oxford University Press, Oxford, p. 29.

Anastas, P. T. and Williamson, T. C., eds. (1996). Green chemistry: An overview, in *Green Chemistry: Designing Chemistry for the Environment*, American Chemical Society Symposium Series No. 626, American Chemical Society, Washington, DC, pp. 1–17.

de Bono, E. (1985). *Six Thinking Hats*, Little, Brown and Company, Boston, MA, p. 205.

Sanderson, K. (2011). It's not easy being green, *Nature*, 469, 18–20.

Tundo, P. and Aricò, F. (2007). Green chemistry on the rise: Thoughts on the short history of the field, *Chemistry International—News Magazine for IUPAC*, 29, September/October, 8pp. http://www.iupac.org/publications/ci/2007/2905/1_tundo.html, accessed on July 4, 2014.

2 Twelve Principles of Green Chemistry

Although some of these principles seem trivial, their combined use frequently requires the redesign of chemical products or processes.

István T. Horváth

LEARNING OBJECTIVES

The learning objectives for this chapter are as follows.

1. Become proficient in recognizing, recalling, and understanding all 12 principles of green chemistry.
2. Become competent in examining the scope of these 12 principles.
3. Learn about atom economy and how it differs from the reaction yield.
4. Practice complex thinking by examining applications of more than one principle of green chemistry in any situation.
5. Start building the basic understanding of toxicity and become familiar with classification of toxic substances.

2.1 INTRODUCTION

The 12 principles of green chemistry have been briefly introduced in Chapter 1 and presented in a simplified form (see Table 1.2). Here, the definitions of these principles are expanded and presented as they were originally proposed by Anastas and Warner (1998). These principles will be discussed further in Chapters 8–13. Here, the scope of these 12 principles is also shown, and examples are provided to illustrate them. In some cases, these principles may remain idealized goals, and innovative efforts of green chemistry may be aimed to realize these goals as close as possible.

2.2 THE 12 PRINCIPLES OF GREEN CHEMISTRY

2.2.1 PRINCIPLE 1: WASTE PREVENTION

This principle was formulated by Anastas and Warner (1998) as follows: "It is better to prevent waste than to treat or clean up waste after it is formed."

Chemical waste is generated virtually in every existing laboratory, such as in the teaching and research laboratories, and in the various industrial settings. If chemical waste is hazardous, its separation from the desired product(s), any further treatment, and the final disposal may require special handling and protective gear and equipment. The latter are often cumbersome to use, are expensive, and may be subject to failure, causing accidents. Hazardous waste is generally costly to dispose of. Also, as discussed in Chapter 1, the waste material that has been disposed of in a presumably safe manner may later be released to the environment via processes that were not initially anticipated. Thus, if there was no hazardous and toxic waste to begin with, it would be the best in terms of health, environment, and economy. Even nonhazardous waste is undesirable, because its separation and removal from the desired product(s) still cost energy and time, and possibly require the use of solvents and other auxiliary substances, some of which may be hazardous.

In some cases, there may be limits to reducing waste, which are inherent to the chemical reaction itself. Here, a typical experiment from the undergraduate organic laboratory is examined.

The objective of the experiment is to prepare methyl *m*-nitrobenzoate from methyl benzoate via nitration with a mixture of concentrated HNO_3 and H_2SO_4. The reaction equation is shown in Figure 2.1.

This is a classical experiment that illustrates the principles of electrophilic aromatic substitution. The electrophile is NO_2^+ (nitronium ion). According to the theory, the predominant product should be the *m* (*meta*) isomer, which is shown in Figure 2.1, rather than the other two possible isomers, namely, *o* (*ortho*) or *p* (*para*) isomers. The structures of all three isomers are shown in Figure 2.2.

This classical experiment is given in many organic laboratory textbooks, which were used by numerous generations. Examples include Fessenden and Fessenden (1983) as an older textbook, and Williamson and Masters (2011) as a recent textbook.

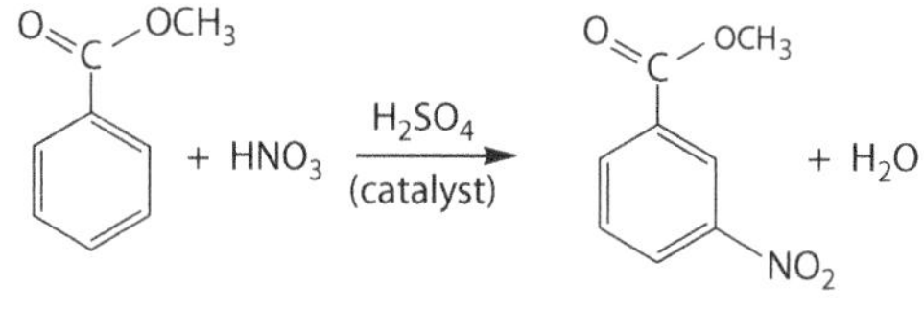

FIGURE 2.1   Nitration of methyl benzoate to give methyl *m*-nitrobenzoate.

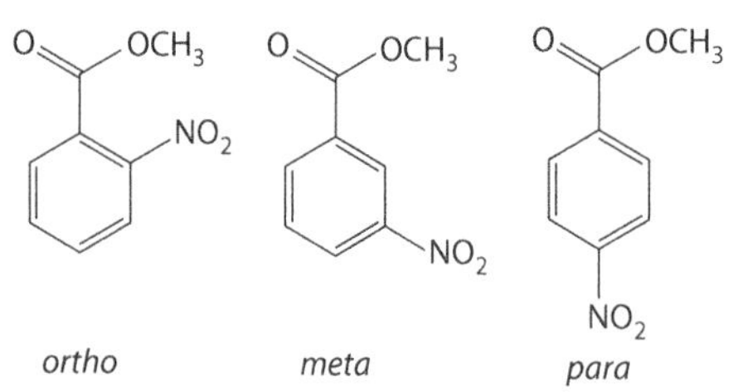

FIGURE 2.2   Three possible isomers of methyl nitrobenzoate.

The reaction indeed gives the *meta* isomer as the predominant product, and this experiment is thus considered the confirmatory one for the theoretical prediction. This is pointed out in both of these textbooks. However, a closer look to the reaction outcome shows that the *meta* isomer is not the exclusive one but is obtained in ~68% yield. The other two isomers are also obtained: the *ortho* isomer in ~28% and the *para* isomer in only a minute amount of ~4% (Fessenden and Fessenden, 1983). These two isomers represent the undesired by-products and are the inevitable waste. This type of waste students cannot control. Other possible organic waste would be the unreacted starting material and di-nitrated product, but this can be controlled by the students by following the experimental protocol carefully. Anastas and Warner point out that the waste that results from the unreacted starting material is costly, because one pays for the substance twice: first as a feedstock and then for its disposal as a waste.

2.2.2　Principle 2: Atom Economy

This principle is given by Anastas and Warner (1998) as follows: "Synthetic methods should be designed to maximize the incorporation of all materials used in the process into the final product."

The concept of "atom economy" is one of the major innovations in the analysis of synthetic efficiency of chemical reactions. It was invented by Barry M. Trost, professor from Stanford University, Stanford, California, in 1991. He then invented many green chemical reactions that are atom economic. The importance of this work was recognized by the Presidential Green Chemistry Challenge Award, which Trost received in 1998.

We can appreciate this innovation much better if we start first with the way success of organic reactions was traditionally evaluated, namely, via chemical yields.

Chemical yield is given as a theoretical and observed yield. Theoretical yield is the maximum amount of product that can be obtained in a reaction, based on the stoichiometry of the reaction. This yield is typically given in grams and is 100% by definition. The observed yield is the actual experimental yield, which is typically reported in the literature as an average of several reaction runs. It is also given in grams and in percentage of the theoretical yield. It is rarely 100%, due to the various factors such as competing reactions, unfavorable equilibrium, incomplete extraction of the products, loss during distillation, recrystallization, or other purification techniques, among others.

The concepts of theoretical and observed yields are illustrated on the reaction of epoxidation of cholesterol with *m*-chloroperbenzoic acid (MCPBA), which is another typical laboratory experiment for beginning organic students. Figure 2.3 shows the chemical equation and the molecular weights (MWs) for the starting materials and products. Figure 2.4 shows calculations of yields for this reaction.

When formulating the principle of atom economy, Trost posed a fundamental question: How much of the reactants end up in the product? This reaction feature is referred to as "atom economy." The general equation for calculations of atom

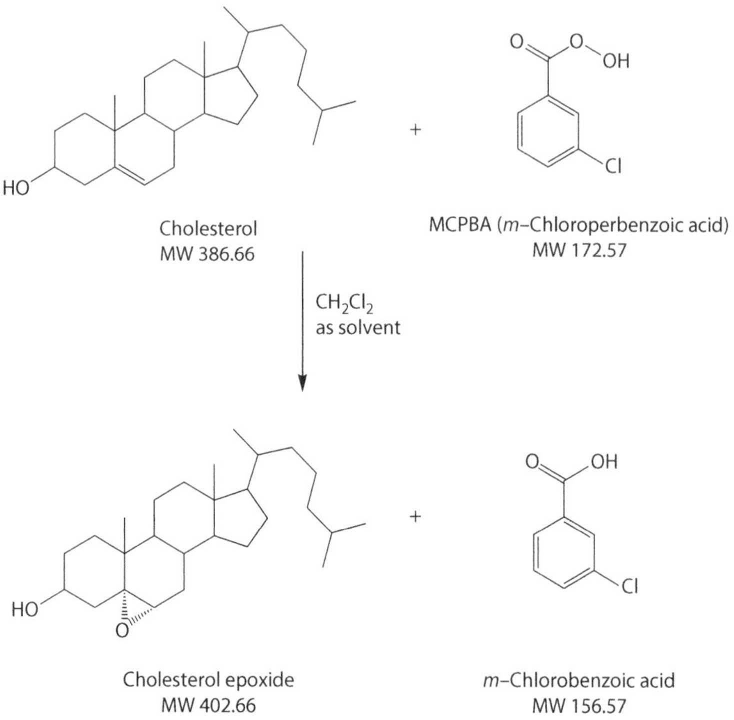

FIGURE 2.3   Epoxidation of cholesterol with MCPBA. MW, molecular weight.

Theoretical yield: 388.66 (g/mol)/402.66 (g/mol) = 0.200 (g)/x

$x = 0.208$ g

Actual observed yield by the student (crude, not recrystallized product) is 0.170 g.

Observed yield (%):

(0.170 g/0.208 g) *100 = 82% ("*" denotes multiplication when "x" may be confused with

an unknown)

FIGURE 2.4   Calculations of yields for epoxidation of cholesterol, if one starts with 0.200 g of cholesterol.

economy and the specific calculations for the epoxidation of cholesterol with MCPBA are shown in Figure 2.5.

We notice that the atom economy of this epoxidation is not very good: only 72%! We can judge qualitatively that atom economy would not be good if we notice that only one atom—oxygen is needed to produce the epoxy product, but the carrier of oxygen, MCPBA, has many more atoms, all of which end up as a waste. The percentage waste in terms of atom economy is 28% (100% − 72%).

General equation:

Atom economy = (Mass of atoms in desired products/Mass of atoms in reactants) × 100

For epoxidation of cholesterol with MCPBA:

Atom economy = (402.66/(386.66 + 172.57)) × 100 = 72.00%

FIGURE 2.5 Calculations of atom economy in general and specifically for epoxidation of cholesterol with MCPBA.

The concept of atom economy enables chemists to quantify the efficiency by which a reaction uses starting materials and the amount of waste produced. Calculations of yields do not show these items.

Exercise 2.1

Calculate the atom economy for epoxidation of cholesterol with magnesium monoperoxyphthalate (MMPP), shown in Figure 2.6.

Answer:

Atom economy = [402.66/(386.66 + 0.5 × 386.55)] × 100 = 69.43%

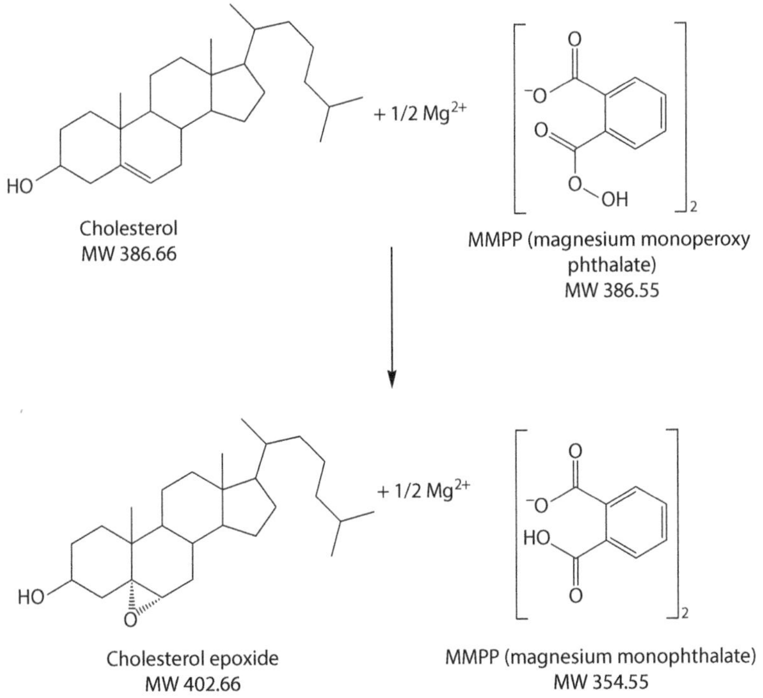

Cholesterol
MW 386.66

MMPP (magnesium monoperoxy phthalate)
MW 386.55

+ 1/2 Mg^{2+}

Cholesterol epoxide
MW 402.66

MMPP (magnesium monophthalate)
MW 354.55

+ 1/2 Mg^{2+}

FIGURE 2.6 Epoxidation of cholesterol with MMPP. MW, molecular weight.

Our calculations show somewhat lower atom economy, 69.43% versus 70.00%, for the epoxidation with MMPP compared to that with MCPBA.

In this case, as in many other cases, the stereochemistry of the product has to be taken into account. In the case of the MCPBA reaction, the epoxide product is obtained intrinsically as a 4:1 mixture of α- and β-isomers (α-isomer: the epoxide ring below the plane, shown as dashed lines in Figure 2.3; β-isomer: the epoxide ring would be above the plane, not shown). This reaction is described in the work of Williamson and Masters (2011, pp. 395–400). However, the MMPP reaction gives pure α-isomer. If our desired product is α-isomer, we should use MMPP reagent.

Exercise 2.2

Calculate the atom economy for the Diels–Alder reaction shown in Figure 2.7.

In this type of reaction, the two starting components (so-called diene and dienophile) add to each other giving a single product (so-called Diels–Alder adduct), and no by-products, which results in a perfect atom economy. The standard calculations show the same thing: Atom economy $= [164.2/(66.1 + 98.1)] \times 100 = 100\%$.

Exercise 2.3

Calculate the atom economy for the Jones oxidation of 2-methylcyclohexanol shown in Figure 2.8.

Calculation of atom economy:

$$\text{Atom economy} = [3 \times 112/(3 \times 114 + 262 + 4 \times 98)] \times 100 = 34\%$$

We notice here an extremely low atom economy for a reaction that typically gives an almost quantitative yield (Nimitz, 1991, pp. 263, 266–267).

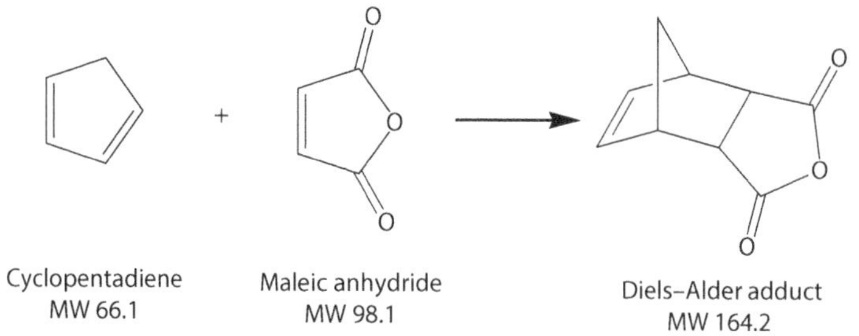

Cyclopentadiene   Maleic anhydride   Diels–Alder adduct
MW 66.1   MW 98.1   MW 164.2

FIGURE 2.7 Diels–Alder reaction between cyclopentadiene and maleic anhydride. MW, molecular weight.

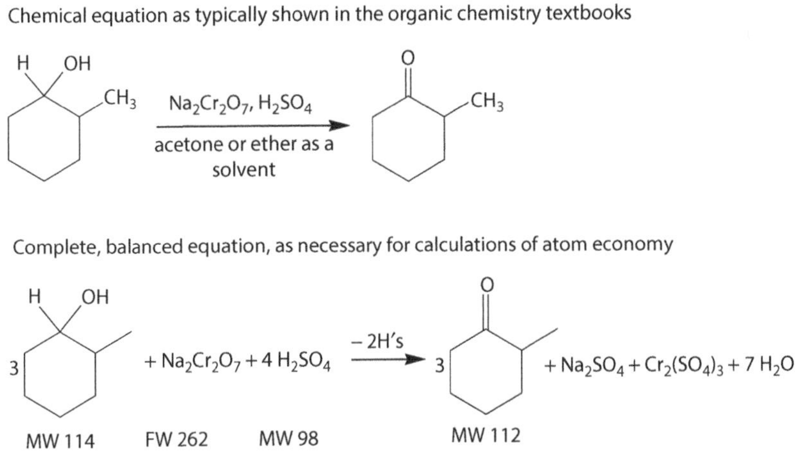

Chemical equation as typically shown in the organic chemistry textbooks

Complete, balanced equation, as necessary for calculations of atom economy

FIGURE 2.8 Jones oxidation of 2-methylcyclohexanol. MW, molecular weight; FW, formula weight.

2.2.3 PRINCIPLE 3: SAFER SYNTHESES

The full description of this principle is given by Anastas and Warner (1998) as follows: "Whenever practicable, synthetic methodologies should be designed to use and generate substances that possess little or no toxicity to human health and environment."

An angle is taken here which is likely to benefit students as they study organic chemistry and perform organic laboratory experiments.

Let us reexamine Figure 2.8, which shows Jones oxidation of a secondary alcohol. The reaction is depicted as customary in organic textbooks. The starting organic materials are on the left of the arrow, and the organic products are on the right. The reagent, in this case an oxidizing reagent, is written above the arrow. Solvents are shown below the arrow. Often, the reagent is represented by a symbol, which in this case would be [O], which stands for oxidation. The usual pedagogical emphasis is on learning the functional group transformations, in this case oxidation of secondary alcohols to ketones.

This approach is not sufficient for green chemistry. One must always consider reagents and their potential hazards. In this case, the chromium reagent is toxic.

The students must learn to connect chemical equation to the corresponding laboratory experiments. A complete, balanced equation, which includes all the reagents, needs to be considered. This is necessary not only for calculations of atom economy, which was shown previously, but also for hazard evaluation. Additional green chemistry metrics exist, which include solvents and environmental factors. These will be covered in Chapters 9 and 10 in conjunction with various industrial applications.

Before the students begin laboratory experiments, they should also make a list of solvents and other substances that they will use during the workup, isolation, and purification of products. These also need to be evaluated for toxicity. Solvents and other auxiliary substances will be further discussed under principle 5. Once all the hazards are evaluated, a brand new design for the reaction, which considers all the elements above, may be proposed by the student, or at least some hazard should be diminished via the appropriate "greening." The students should not assume that all of the experiments in their laboratory manual are either green or optimized for green principles. Thus, the exercise on new design and greening is real, and not just a make-work project.

2.2.4 PRINCIPLE 4: SAFER PRODUCTS

The full description of this principle is given by Anastas and Warner (1998) as follows: "Chemical products should be designed to preserve efficacy of function while reducing toxicity."

Implementation of this principle requires additional knowledge about toxicity, which the chemistry students typically do not get from their regular chemistry courses. To partially remedy this, the types of toxicity, and some basic terms and definitions about toxic effects are presented in Table 2.1. Some examples of toxic chemicals are given in Table 2.2. The structures of the toxic compounds from Table 2.2 are shown in Figure 2.9.

Thanks to the advances in chemistry and related sciences, we now understand better the relationship between the chemical structure and the resulting properties of the products. Examples include color (of dyes and paints), amphiphilicity (property required for detergent-type action), tensile strength (of various fibers),

---

TABLE 2.1
Types of Toxicity and Some Basic Terms and Definitions about Toxic Effect

Carcinogens: Agents involved directly in causing cancer

Chemical toxins: Inorganic (mercury, lead), organic (benzene, vinyl chloride)

Classification of carcinogens: Definitely carcinogenic to humans, probably carcinogenic to humans, suspected human carcinogens, and confirmed animal carcinogens with unknown relevance to humans, among others. Classification differs somewhat among different agencies, such as the International Agency for Research on Cancer, GHS, the US National Toxicology Program, and the American Conference of Governmental Industrial Hygienists, among others

Co-carcinogens: Promote the activity of carcinogens (but do not necessarily cause cancer on their own)

Developmental toxins: Cause adverse effects in the development at any stage of life and may include those induced during pregnancy

Mutagens: Agents that change the genetic material; they can also be carcinogenic, but not necessarily

Pro-carcinogens: Precursors to carcinogens (they turn to carcinogens in the body, e.g., but are not carcinogenic by themselves)

Pro-mutagens: Form mutagens (e.g., via cellular metabolic processes), but are not mutagenic by themselves)

Reproductive toxins: Interfere with normal reproduction by causing adverse effects on fertility, development of the offspring, and birth defects (in which case they are called teratogens), among others. Endocrine disruptors have received much attention recently

Toxic effects: Dose dependent, species specific, organ specific

Toxicity: Chemical, biological, physical (only chemical will be considered further)

Toxicity influenced by: Acute exposure, chronic exposure

TABLE 2.2
Examples of Toxic Chemicals

Carcinogens	Mutagens	Teratogens/Reproductive Toxins
Benzene, an industrial solvent, now largely replaced with less toxic alternatives	5-Bromouracil, a nucleobase analog	Diethylstilbestrol, an artificial estrogen
2-Naphthylamine, used to make azo dyes	Tris(2,3-dibromopropylphosphate), "tris," a flame retardant in textiles and plastics, now largely replaced with less toxic alternatives	Thalidomide, R-isomer, used in the past against nausea in pregnant women
Vinyl chloride, monomer, used for production of polyvinyl chloride	Hydroxyl amine, a reagent	13-*cis*-Retinoic acid, "Accutane," used for treatment of acne
N-Nitrosodimethylamine, an industrial by-product		

Note: There may be an overlap between categories.

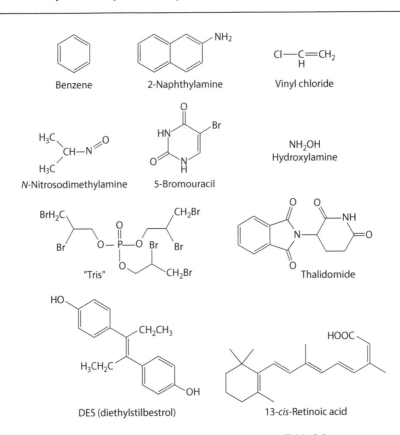

FIGURE 2.9 Structures for the compounds given by name in Table 2.2.

and nonflammability (of various fabrics). When designing various pharmaceuticals, medicinal chemists and pharmacologists use the so-called structure–activity relationships, which correlate selected parts of the chemical structure, including molecular shapes, to a particular biological function. The latter function is often elicited by binding the drugs to their specific receptors. This is where the knowledge of stereochemistry, a branch of organic chemistry, will become especially useful, because the drug–receptor fit is often like a key fitting the lock, and thus, three-dimensional and chiral (handedness) properties are essential.

To design chemicals that are not toxic, we need to learn as much as possible about the reasons why some chemicals are toxic and some others are not, and why some species are affected more by a particular toxic chemical than some others. We need to be able to predict in advance, as much as we can, if the chemical we intend to make will be toxic or not. Further, the dose of chemicals to which a living entity is exposed matters to the toxic effect. Something may not be toxic in small amounts, but, just like in a case of a drug overdose, may be quite toxic if the amount is above some limit. Some chemicals accumulate in the fatty tissues in the body, and the knowledge of partition coefficient (distribution between lipid and water) will be needed. We also need to be able to sort out all of these factors, and there are many more. This may appear to be a formidable task, but we shall conquer it gradually.

Two exciting contemporary developments in the field of designing nontoxic chemicals are the U.S. Environmental Protection Agency (EPA)'s Computational Toxicology Research Program and ToxCast, a high-throughput screening project. In the latter, biochemical assays, such as binding to the cellular receptors, are applied to chemicals for which toxicology data are already available. Based on these data, statistical and computational models are built for prediction of toxicity based on the assays alone (Sanderson, 2011). The EPA's website (http://www.epa.gov/ncct) describes the above initiative. Students and other readers are encouraged to explore the content of this site.

2.2.5 PRINCIPLE 5: SAFER AUXILIARIES

This principle is given by Anastas and Warner (1998) as follows: "The use of auxiliary substances (e.g. solvents, separation agents) should be made unnecessary whenever possible and innocuous when used."

Auxiliary substances are not the starting materials, reagents, catalysts, or products, and thus are not a part of the chemical equations. Instead, these are substances that are used during the workup, separation, isolation, and purification of the products. They are routinely used in syntheses in the laboratory, manufacturing, and processing or chemicals for various uses.

Various solvents are notorious for being harmful, especially halogenated solvents. Many of them are hazardous to health or environment. Examples are given in Table 2.3.

One of the desired properties of solvents that are used for isolation of products, their purification, and manipulation is low volatility. This property makes them easy to remove at the end of these procedures. However, the low volatility may carry

TABLE 2.3
Examples of Hazardous Halogenated Solvents

Solvent	Hazard
CH_2Cl_2 (methylene chloride)	Suspected carcinogen (among other adverse effects on health)
$CHCl_3$ (chloroform)	Suspected carcinogen (among other adverse effects on health)
CCl_4 (carbon tetrachloride of carbon "tet")	Suspected carcinogen (among other adverse effects on health)
CFCs (chlorofluorocarbons)	They deplete the stratospheric ozone layer, thus environmental hazard

other hazards, which are in common with all volatile organic compounds. Volatile compounds in general can escape easier from their containers, and this potentiates their health hazards, flammability, and negative effect on the environment such as smog formation.

The green chemistry initiatives are to either eliminate solvents altogether in the so-called solventless reactions or use benign solvents, such as water. These initiatives will be discussed in detail throughout other chapters, notably in Chapters 4 through 7.

2.2.6 Principle 6: Energy Efficiency

Anastas and Warner (1998) give the full description of this principle as follows: "Energy requirements should be recognized for their environmental and economic impacts and should be minimized."

Energy consumption is involved in virtually all stages of chemical experiments, manufacturing, and processing. Examples that students can relate to are heating or cooling of the reaction mixtures, heating required for distillation of products and solvents, melting, sublimation, and recrystallization processes, among others. On the industrial scale, energy requirements are amplified and often require costly equipment to safely dissipate heat to avoid overheating and runaway reactions. Many energy uses are assumed to be necessary, but they may not be, and thus, the greening can often be done by saving energy. One example is heating under reflux. It is a common laboratory procedure, which is almost universally used, because it results in heating at a constant temperature. As the reaction mixture boils, its vapors are returned to the reaction flask by using a vertical cooled condenser. This is shown in Figure 2.10.

In most of the cases, the heating requirements have not been determined. Thus, the energy requirement has not been optimized. It is possible that in many cases, comparable results could be achieved at temperatures that are lower than the reflux one. A reasonable compromise about the reaction time, which would perhaps need to be longer at lower temperature, needs to be considered. One could use a setup shown in Figure 2.11, for heating at a constant temperature.

Energy saving may also be accomplished by using microwave heating, when feasible.

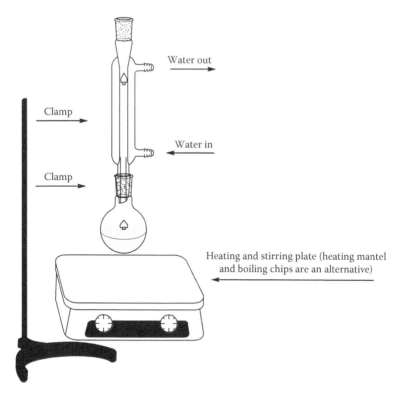

FIGURE 2.10   A setup for heating under reflux.

2.2.7 PRINCIPLE 7: RENEWABLE FEEDSTOCKS

The full definition of this principle, as given by Anastas and Warner (1998), is as follows: "A raw material or feedstock should be renewable rather than depleting, wherever technically and economically practicable."

Fossil fuels belong to the resources that can be depleted, whereas renewable feedstocks are typically associated with biological materials, such as plant-based materials.

This principle is strongly correlated with sustainability. As we utilize the resources that are available to us, we need to find a way to have such resources available also for future generations.

We need to switch from petroleum hydrocarbon resources to other resources, because they are not green in many respects. As one example, petroleum hydrocarbons are converted to other useful products typically by oxidation, which often requires catalysis by metals, such as chromium, which are toxic. We see in this example that the principles of green chemistry are often intertwined and are difficult to be looked at in isolation.

The use of biological feedstock is promising. It also has a requirement to be renewable. Much of the green chemistry innovation is in this area, which we shall address in Chapters 8 and 9.

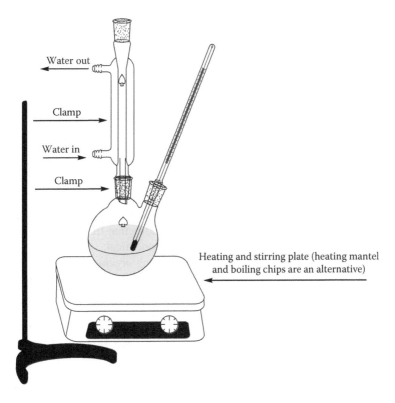

FIGURE 2.11   A setup for heating at constant temperature.

2.2.8   PRINCIPLE 8: DERIVATIVE REDUCTION

Anastas and Warner (1998) give the definition of this principle as follows: "Unnecessary derivatization (blocking group, protection/deprotection, temporary modification of physical/chemical processes) should be avoided whenever possible."

The derivatization principle is illustrated on an example from the beginning of organic chemistry. Starting from compound **1**, which has a keto and ester groups, we need to make compound **2**, in which the ester group is converted to an alcohol, whereas the keto group is preserved. This desired transformation is shown in Figure 2.12.

The reaction requires a reduction of the ester group to an alcohol. Two reducing agents that students are generally familiar with are sodium borohydride, $NaBH_4$, and lithium aluminum hydride, $LiAlH_4$. The latter will reduce the ester group, whereas the former will not. However, both reducing agents will reduce the keto group to an

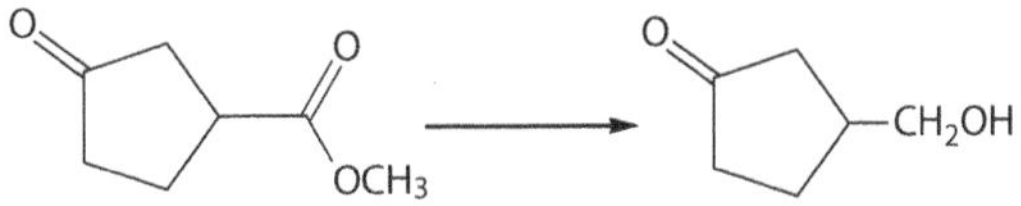

1 a keto ester   **2** a keto alcohol

FIGURE 2.12   Transformation of a keto ester **1** to a keto alcohol **2**.

alcohol, because the keto function is more easily reduced than the ester. To solve this synthetic problem, one can react **1** with ethylene glycol, **3**, under the condition of acid catalysis, to give compound **4**, in which the keto group is transformed to the acetal **5**, whereas the ester group is unchanged (it does not react with **3**). One can then reduce the ester group in **5** with LiAlH₄, which will convert it to the alcohol **6**, while leaving the acetal group intact. Finally, the acetal group is removed by an acid hydrolysis to give the desired product **2**. The terms chemists use are that the keto group is "protected" against reduction by "derivatization" into an acetal, which is later removed to "deprotect" the ketone. This sequence of protection, reduction, and deprotection is shown in Figure 2.13.

In this example, and at the level of a beginning organic student, the protection/deprotection was necessary. A quick examination of the reaction sequence reveals a poor atom economy, because ethylene glycol will go to waste. Alternatively, a possible recycling of ethylene glycol would require extraction from the aqueous waste and purification for its reuse, which may not be economical.

Another type of derivatization that the beginning organic students are familiar with is the introduction of a functional group that renders the molecule more reactive, and then replacement of this group is carried out with another group, which is desired. For example, we wish to convert a primary alcohol R-CH₂OH, **6**, to the corresponding chloride, R-CH₂Cl, **7**, via a nucleophilic substitution S_N2, with Cl⁻ ion. The reaction will fail, because OH⁻ is a poor leaving group. However, if we derivatize the alcohol, by converting the OH group into a good leaving group such as tosylate, the reaction will occur. This process is shown in Figure 2.14.

In this example, we see also a poor atom economy, because tosylate (TsO⁻) will go to waste.

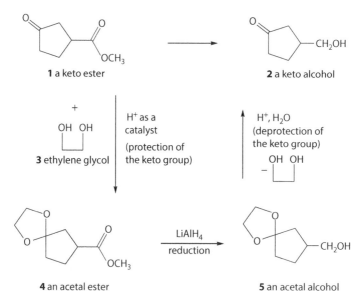

FIGURE 2.13 Transformation of a keto ester **1** to a keto alcohol **2** via a protection/deprotection scheme.

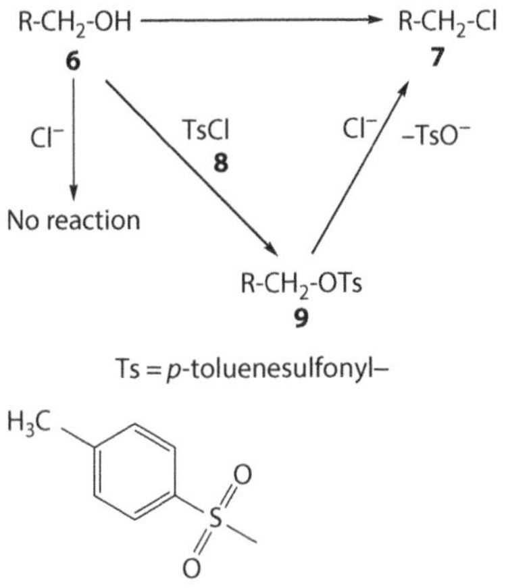

FIGURE 2.14 Derivatization to facilitate the substitution reaction.

2.2.9 PRINCIPLE 9: CATALYSIS

Anastas and Warner (1998) give the definition of this principle as follows: "Catalytic reagents (as selective as possible) are superior to stoichiometric reagents."

Catalysts typically lower the activation energy barrier for the reaction, thus enabling reactions to occur faster and become feasible at the lower temperatures. This translates into the greener energy requirements.

Many catalysts have been developed to either enable or enhance selectivity of the reaction. Stereoselectivity and achievement of enantiomeric excess are necessary in the syntheses of fine chemicals, especially drugs. Enantiomers exhibit the opposite handedness, like the left and the right hand. Often only one of the enantiomers has the desired biological properties, and the other one may either be toxic or place burden on the organism's detoxification system. The production of the undesired enantiomers creates waste, but the more serious problem is the required separation of enantiomers for the purpose of isolating the biologically active one. This process is often difficult, lengthy, and expensive. It is often a major factor that drives up the cost of drugs. It is therefore very important that the chemical synthesis yields the desired enantiomer either exclusively or in excess. This is often achieved with special catalysts. We shall address such catalysts in Chapters 9 and 10, especially on the examples of green pharmaceutical syntheses.

Selectivity of catalysts is useful for many basic chemical reactions that are familiar to students. One such example is shown in Figure 2.15.

When alkyne **10** reacts with H_2 in the presence of Pt, the reaction does not stop at the alkene stage, but gives the alkane **11**, a completely saturated product. If less than two equivalents of H_2 were used, one would still get the alkane, but would have some unreacted **11**. No alkene would be obtained. In contrast, when a less reactive catalyst is used, such as Ni_2B ("P-2"), the reaction stops at the alkene stage. In both cases, the hydrogenation takes place at the surface of the catalyst. This accounts for the *cis* stereochemistry of **12**.

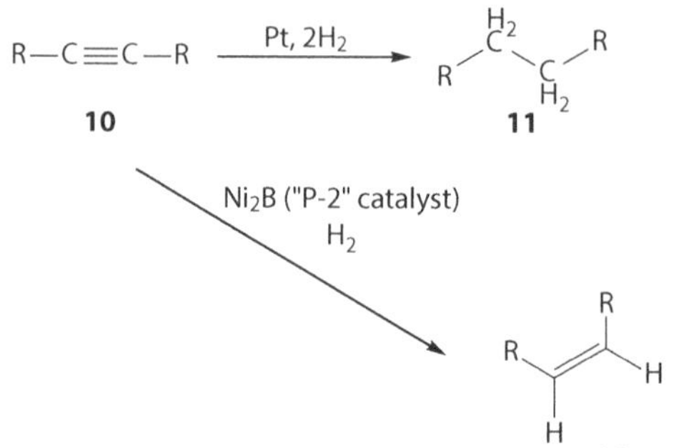

FIGURE 2.15   Selectivity in hydrogenation of unsaturated hydrocarbons.

Catalysis in general offers advantages over typical stoichiometric reactions, because many of them need to be facilitated, often by adding an excess of starting materials or additional reagents, which drive the cost up and typically end up in waste.

2.2.10   PRINCIPLE 10: DEGRADABILITY

The full description of this principle, as given by Anastas and Warner (1998), is as follows: "Chemical products should be designed so that at the end of their function they do not persist in the environment and break down into innocuous degradation products."

Many useful products, such as various plastics, have been found to be persistent in the environment. This has created multiple problems, from overflowing landfills, to the danger for wild life, such as sea birds, which ingest these products, often with consequences that are detrimental for their health. For such reasons, chemists started creating plastics that can be degraded in the environment, often with the help of various microbes (thus the term "biodegradable").

Some other useful products, such as pesticides, tend to accumulate in living organisms (thus the term "bioaccumulation"), especially in the lipid tissues. The organisms are thus exposed to such chemicals over a prolonged period of time, which can exacerbate toxicity. It can also lead to the transfer of such chemicals from mothers to their nursing infants via milk. Degradability in cases of pesticides is therefore a desired property.

Degradation of chemical products should give benign end products. The design of chemical products needs to fulfill the degradability requirement. Green chemistry considers chemical degradation that often occurs in the environment via hydrolysis, photolysis, and oxidation, among other means. Also, various microbes and plants may be used to remove/degrade toxic chemicals.

2.2.11   PRINCIPLE 11: POLLUTION PREVENTION

The short definition of this principle was given by Sanderson (2011) and was further expanded on as follows: "Develop methods for real-time monitoring and control of chemical processes that might form hazardous substances." Anastas and Warner (1998) give the definition of this principle as follows: "Analytical methodologies need to be further developed to allow for real-time, in-process monitoring, and control prior to the formation of hazardous substances."

This principle focuses on the development of accurate and reliable analytical techniques that can be used to monitor the generation of hazardous by-products and occurrences of side reactions. Once these are observed, parameters of such reactions and processes can be adjusted to eliminate or reduce their formation. Especially important is the determination of the completion level of the chemical reactions, so that one can intervene before quenching them. One simple example, which is familiar to most of students, is to follow the reaction progress by thin-layer chromatography. If the starting material is still present, the reaction time is increased to allow for all the materials to react. Modern analytical techniques are capable of detecting minute amounts of hazardous or undesirable chemicals. Technological capabilities exist for interfacing analytical sensors within process control, and thus minimizing hazards in an automated fashion.

2.2.12 PRINCIPLE 12: ACCIDENT PREVENTION

This principle is given by Anastas and Warner (1998) as follows: "Substances and the form of a substance used in a chemical process should be chosen so as to minimize the potential for chemical accidents, including releases, explosions, and fires."

This is an important principle that broadens the scope of green chemistry to prevention/minimization of hazards above and beyond toxicity. Other hazardous properties, such as flammability and tendency to explode, are often responsible for injuries and large-scale releases of chemicals into the environment.

It is important to evaluate some other green chemistry principles in light of this one. For example, recycling of solvents is green according to principle 7, but solvents are often flammable and pose fire or explosion hazards.

One green technology is a rapid consumption of hazardous substances, so that they do not need to be stored.

REVIEW QUESTIONS

2.1 Research information about toxicity and other possible hazardous properties of the starting materials and products for the following reactions that are given in this chapter:
a. Nitration of methyl benzoate
b. Epoxidation of cholesterol with MCPBA
Search the Internet by the chemical names. Look for the material safety data sheets (MSDSs) and globally harmonized system (GHS) data for these chemicals, which contain information about toxicity and hazardous properties. Also look up Wikipedia information, ToxNet (http://toxnet.nlm.nih.gov/), PubChem (http://pubchem.ncbi.nlm.nih.gov/ and https://www.osha.gov/dsg/hazcom/ghs.html [a guide to the GHS of classification and labeling of chemicals], and http://www.sigmaaldrich.com/

safety-center/globally-harmonized.html), among many other available sources.

2.2 Prepare index cards with the 12 principles of green chemistry, with the complete definitions of the principles, as given by Anastas and Warner. Mix up the cards and randomly select four cards. Try to draw a correlation between the principles. Especially look for the ways in which these principles may reinforce each other. Repeat this exercise several times, with a new combination of cards.

ANSWERS TO REVIEW QUESTIONS

2.1 (a and b) Detailed answers can be found in the suggested resources, especially MSDS/GHS. Examples of MSDS from a variety of sources, including suppliers for these chemicals: for methyl benzoate—http://www.sciencelab.com/msds.php?msdsId=9927228; for *m*-nitromethylbenzoate (may be listed under its alternative name, methyl 3-nitrobenzoate)—http://www.coleparmer.com/Assets/Msds/03248.htm; for cholesterol—http://www.reflec.ameslab.gov/docs/Cholesterol.htm; and for cholesterol epoxide—http://datasheets.scbt.com/sc-214687.pdf.

The MSDS/GHS data show that not all the toxic effects are known in all the cases. Hazard is often considered for large-scale exposure, such as in the production or processing plant. In student laboratories, the amounts of materials that are used are very small.

2.2 Answers will vary, depending on the choice of cards. This exercise is also useful as a group practice in class.

REFERENCES

Anastas, P. T. and Warner, J. C. (1998). *Green Chemistry: Theory and Practice*, Oxford University Press, Oxford.

Fessenden, R. J. and Fessenden, J. S. (1983). *Techniques and Experiments for Organic Chemistry*, PWS Publishers, Willard Grant Press, Boston, MA, pp. 262–263, 269–273, for nitration of methyl benzoate.

Horváth, I. T. (2002). Green chemistry, *Acct. Chem. Res.*, 35, 685.

Nimitz, J. S. (1991). *Experiments in Organic Chemistry: From Microscale to Macroscale*, Prentice Hall, Englewood Cliffs, NJ, pp. 263, 266–267, for oxidation of 2-methylxyclohexanol with Jones reagent.

Sanderson, K. (2011). It's not easy being green, *Nature*, 469, 18–20.

Trost, B. M. (1991). The atom economy—A search for synthetic efficiency, *Science*, 254, 1471–1477.

Trost, B. M. (2002). On inventing reactions for atom economy, *Acct. Chem. Res.*, 35, 695–705.

Williamson, K. L. and Masters, K. M. (2011). *Macroscale and Microscale Organic Experiments*, 6th edn., Thomson Brooks/Cole, Belmont, CA, pp. 395–400 for epoxidation of cholesterol; pp. 401–405 for nitration of methyl benzoate.

3 Innovative Aspects of Green Chemistry

Incredibility escapes recognition.

Heraclitus
The Fragment LXXXVI

LEARNING OBJECTIVES

The learning objectives for this chapter are as follows.

Objective 1. Learn about innovation in general	*Objective 2.* Learn about some common thinking types and their potential for fostering innovation	*Objective 3.* Learn about some specific innovations in green chemistry
Innovation in general and in green chemistry (3.1)	Deductive thinking (3.2.1)	Reactions in water (3.3.1)
Does innovation require a special type of thinking? To answer this, move to Objective 2.	Inductive thinking (3.2.2)	Reactions in superheated water (3.3.2)
	Critical thinking (3.2.3)	Reactions in water in nanomicelles (3.3.3)
	Linear and nonlinear thinking (3.2.4)	Design of halogen-free firefighting foam (3.3.4)
	Lateral and vertical thinking (3.2.5)	Sources of examples of innovation in green chemistry (3.3.5)
	Complex thinking (3.2.6)	

Note: Section numbers are in parentheses.

3.1 ABOUT INNOVATION IN GENERAL AND IN GREEN CHEMISTRY

The common dictionary meaning of "innovation" is the introduction of something new, such as ideas, devices, methods, or processes. Innovation usually refers to the implementation and application of novel idea, method, and so on, whereas invention typically describes the creation of these (see Wikipedia). According to Padgett and

Powell (2012), something is novel if it does not exist in the current practice and is not anticipated. Novel works are those not obvious to others skilled in the same field.

There are also business definitions of innovation (see web resources listed in the references). These definitions are suitable for describing innovations in green chemistry either in the form in which they are stated or with some modifications. For example, innovation is defined as follows: "The process through which economic and social value is extracted from knowledge through the generation, development, and implementation of ideas to produce new or improved strategies, capabilities, products, services, or processes" (Conference Board website, www.conferenceboard.ca/cbi/innovation. aspx), and "To be called an innovation, an idea must be replicable at an economic cost and must specify a specific need" (Business dictionary website, www.businessdictionary. com/definition/innovation.html). These definitions/descriptions resonate well with the principles of green chemistry, which include economic aspects either implicitly or explicitly (as in principles 6 and 7).

Some innovations in green chemistry are general, such as a focus on prevention rather than protection. Some others are specific for a particular chemical reaction of process. These involve both the design of new green reactions and the greening of existing ones.

Importantly, green chemistry draws much of its innovation from the existing knowledge, which has not been previously recognized to have practical applications (e.g., organic reactions in aqueous suspensions).

The design of green reactions consists of several steps. First, chemists write down the synthetic reaction sequence on paper. This step is traditionally termed "paper chemistry" (although it can be done on computer!). To verify the proposed reaction in the laboratory, chemists first need to devise the reaction protocol. The latter consists of detailed instructions on executing the reaction in the laboratory, and on the isolation and identification of products. Typically, the protocol describes the equipment, reaction conditions (heating, cooling, etc.), quenching the reaction (with water, acid, or base, for example), the workup (e.g., extraction with solvents), isolation (e.g., removing the solvent), purification (such as crystallization or distillation), analytical procedures for monitoring the reaction progress and determining the purity of the product (e.g., by chromatography, spectroscopy), and structural identification (e.g., by spectroscopy). After the protocol is established, the reaction is run in the laboratory following the protocol. If the proposed reaction does not work in the laboratory or gives poor results, chemists modify both the reaction steps and the protocol until the reaction is realized in practice. The reaction may still need to be optimized to give the maximum greenness. Often, the devil is in the details. Innovation may be required in execution of the seemingly minor aspects of the original design and protocol, but without which the implementation of the reaction in practice may fail.

The task of designing a novel green reaction as a series of steps is shown in Table 3.1.

In practice, the overall design requires going back and forth between the steps, changing, adjusting, and tweaking them. Although steps 1–3 are applicable to the development of any novel reaction, for a novel green reaction we need to consider also step 4 and include the 12 principles of green chemistry in all the steps. What appeared to be a relatively straightforward stepwise process, moving from

TABLE 3.1

Design of a Novel Green Reaction

1. Design the reaction "on paper."
2. Design the laboratory protocol.
3. Verify if the reaction works in the laboratory, by following the protocol. If not, modify the previous steps as necessary.
4. Evaluate if the reaction and the protocol are as green as proposed; if necessary, optimize the greenness of the steps.

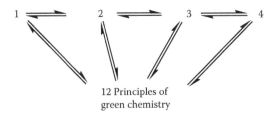

12 Principles of
green chemistry

FIGURE 3.1 Interconnections between the steps for the design of a novel green reaction with inclusion of 12 principles of green chemistry. See Table 3.1 for the steps.

step 1 to step 4, now becomes a rather elaborate interconnected system. This is shown in Figure 3.1.

The scheme shown in Figure 3.1 is simplified. The interconnections between the individual 12 principles of green chemistry are not shown. In a real design, we need to include interactions of these principles with each other and also with each of steps 1 through 4. Showing all of these interactions in the scheme would make it extremely complicated, and it may obscure the steps of the process.

A question arises: How can we think about the necessary steps and their interconnections at the same time, and how can we organize our thinking process to avoid frustration and to achieve success? Further, is there a thinking method that will help us innovate, or is innovation some sort of inspiration, which may or may not happen? Selected materials on thinking methods, which should help students navigate through innovating aspects of green chemistry and build up their own innovative capabilities, are presented in Section 3.2.

3.2 ABOUT THINKING IN GENERAL AND FOR GREEN CHEMISTRY INNOVATIONS

In this section, we briefly summarize various ways of thinking and reasoning in a simplified manner but provide some useful references for the reader. We point out various reasoning errors, which can derail our thinking about green chemistry. Finally, we show that all types of thinking need to be integrated and used in an inclusive way to be able to tackle the complex problems found in green chemistry.

3.2.1   DEDUCTIVE THINKING

Deductive thinking is commonly used, especially in the sciences and mathematics. This type of thinking was formulated by Aristotle (Lawhead, 2001) and is based on his logic rules. One starts from the accepted premise, such as a general principle assumed to be true. One then examines a specific case, following the rules of logic in a stepwise manner. One concludes that something must be true if it is a special case of the general principle that is known to be true. An example of a deductive inference is that if all birds lay eggs, then pigeons must lay them too, because pigeons are birds (Reif, 2010, p. 111). Reif gives an example from mathematics. The number π is defined as the ratio of the circumference C of a circle compared to its diameter D, namely, $\pi = C/D$. This definition is the starting point in our reasoning, and we accept it as true without questioning it. Then a simple deductive inference gives a conclusion that a circumference of a particular circle can be obtained by multiplying its diameter by π. This reasoning is presented by an equation, $C = \pi D$. Likewise, $D = C/\pi$. Reif reminds us that it would be foolish to remember these conclusions as separate facts when they can be easily inferred from the definition of π. The deductive method can thus be looked at as formulaic (algorithmic). We give another familiar example of deductive inference, starting from an equation that defines force as $F = ma$ (force is mass times acceleration). If we plug in specific values for mass and acceleration into this equation, we should obtain the value for the force in a predictable and reliable manner, and can be confident that the answer is correct. If force is equal to mass times acceleration in general, then this must hold for the specific cases of mass and acceleration. Alternatively, if we know, for example, force and acceleration, we can obtain mass from the same equation.

The major shortcoming of deductive reasoning is that it is based on premises that are not questioned. If the premise is incorrect, following the rules of logic will provide a logically correct answer, which, however, will be a wrong answer!

Examination and questioning of already accepted premises may lead to innovation. The green chemistry movement examined and rejected the premise that protection is the exclusive way of dealing with the problem of toxic chemicals. The protection premise was wrong, and its substitution with a new premise, that of prevention, was innovative.

3.2.2   INDUCTIVE THINKING

The weakness of the deductive reasoning, that it becomes a house of cards if the original premises are incorrect, was pointed out by Francis Bacon, who formulated a different way of thinking, that of induction (Lawhead, 2001). Inferring by induction starts with a limited knowledge of a few specific cases and arrives at plausible conclusions. The process of induction involves making specific observations; looking for similarities, differences, and possible analogies to what is already known; identifying patterns and regularities; examining quantitative relationships and comparing them with the known phenomena, and finally formulating a hypothesis. However, Bacon's system also becomes a house of cards if the supposedly unrelated phenomena are considered related. Inferring by analogy involves observing that two systems have some similar features, and then concluding that other properties of the two systems are probably also similar. This may lead to a wrong conclusion (Reif, 2010, pp. 110–110). Nickerson et al. (1985) list various specific instances of making errors in induction: failing to sample enough information, failing to sample in an unbiased way, ignoring negative information or not exploiting it, failing to discard and clinging to the unlikely hypotheses, considering too few alternatives, and failing to view one's own opinion or hypothesis objectively, among others.

Novelty can come from identification and careful examination of reasoning errors. We give here one example from organic chemistry. Numerous observations have shown that organic chemicals that are water insoluble react in organic solvents. An inductive generalization was made that organic chemicals that are insoluble in water necessitated that their reactions be run in organic solvents. This was based on well-known solubility rules and the kinetic requirement that reactant molecules need to collide in order to react. Some isolated examples of organic materials that are not water soluble but still react in water existed, but were not included in the reasoning, which was erroneous. Green chemistry examined these isolated reactions and found them to be representative of a much larger class of similar reactions. Thus, previously stated generalization that one needs organic solvent was not correct. Green chemistry exploited its discovery and started an entirely new subfield of organic reactions in water.

3.2.3 CRITICAL THINKING

Critical thinking has various definitions and numerous attributes, which is not surprising, because it is a subject of an entire relatively recent educational movement. We select here some definitions and descriptions that are representative and are useful for green chemistry applications.

Critical thinking is used to decide if a claim is true, partially true or false. It is concerned primarily with judging the true value of statements and is seeking errors. It is thinking that is rational, open-minded, and informed by evidence. It results in a judgment by analyzing and evaluating relevant criteria. Critical thinking calls for the ability to reconstruct one's beliefs on the basis of new evidence (see Wikipedia).

Critical thinking is necessary but not a sufficient condition for creativity. However, when creativity generates novel possibilities, we need critical thinking to evaluate them (Nickerson et al., 1985, pp. 88–89).

Critical thinking helps us reevaluate the accepted paradigms. This can lead to innovation, as shown previously on the example of green chemistry. We add here another important feature of critical thinking that is essential for green chemistry innovation, namely, the ability to reconstruct one's belief based on new evidence. Just because things were done in a certain way in the past and were justified by then available evidence, it does not mean that we have to cling to these ways. We should constantly reevaluate the ways based on new evidence. For example, toxic solvents were once thought to be unavoidable, but the new evidence showed that water, the ultimate benign solvent, can be used in many cases.

3.2.4 LINEAR AND NONLINEAR THINKING

Again we give a simplified description, which we believe will be useful for green chemistry. Terms "linear thinking" and "nonlinear thinking" have become popular relatively recently, especially in the field of management (see, e.g., Osterman et al., 2013). These two types of thinking show different preference of methods for comprehending information. Linear thinking is based on rationality, logic, and analytical reasoning, whereas nonlinear thinking utilizes preferentially intuition, insight and creativity. Linear thinking is valued for its efficiency. However, it is prone to the same problems as deductive thinking, because both depend on logic. Linear thinking is not well suited for solving problems in complex systems in which various parts are interconnected and influence each other. Although we can apply linear thinking to a single connection within a complex system, this does not guarantee success and often leads to a dead end. The sum of the individual connections often does not represent the system adequately, because different parts of the system may be changed by the interactions with other parts, and thus cannot be considered in isolation.

One example is the application of 12 principles of green chemistry to the greening of a chemical system. If we think in a linear fashion, we shall focus only on one principle and will try to do green optimization based only on this selected principle. For example, we choose to focus only on atom economy, but do not include other principles, such as economic factors. By following the linear path, we may design a novel reaction with high atom economy, and thus no waste. However, if we did not consider economic factors, the reaction may be too expensive to be implemented in practice.

Nonlinear thinking uses multiple perspectives and often examines connections that are not immediately obvious. This approach may lead to novel and unexpected solutions to the problem. A demand for this type of thinking style is growing, especially in business, management, and the workplace in general.

Multiple perspectives are often obtained from interdisciplinary approaches. We have already seen that looking at green chemistry reactions and processes, we need to look at them also from the points of view of toxicology, environmental science, and economy, for example.

Both linear and nonlinear thinking should be practiced. We should strive to utilize the best aspects of each as appropriate for the problem that we are examining. The same is true for deductive and inductive reasoning.

3.2.5 LATERAL AND VERTICAL THINKING

The terms "lateral thinking" and "vertical thinking" were coined by Edward de Bono. Vertical thinking is traditional thinking in which one takes a position and then builds on the basis of that position, in a stepwise and logical manner. Vertical, deductive, and linear thinking are related because they share the tools of logic. In lateral thinking, we search for different approaches and different ways of looking at things, and we move "sideways" to perceive things differently and to try different concepts and points of entry (de Bono, 1993, p. 53). Lateral thinking challenges assumptions and generates alternative ways of looking at the problem. It breaks out of the concept prison of old ideas (de Bono, 1990, p. 11). It is involved in creating something new, rather than analyzing something old. It explores multiple possibilities and approaches, instead of pursuing a single approach. Lateral thinking is creative and leads to innovation. This type of thinking is best captured by de Bono's description: "You cannot dig a hole in a different place by digging the same hole deeper" (de Bono, 1993, pp. 52–53). Lateral thinking corresponds to digging the hole in a different place, whereas vertical thinking corresponds to digging the hole in the same place.

Green chemistry can be well described by "digging the hole" in a different place, that of prevention, whereas traditional chemistry just "dug the hole" in the same place, that of protection, but deeper and deeper.

3.2.6   COMPLEX THINKING

The term "complex thinking" has been used in different contexts. We present here a simplified coverage of this topic and pick and choose some elements that are useful for green chemistry.

Complex thinking combines various aspects of thinking, among which are the following four: problem solving, critical thinking, creative thinking, and finally decision making, in which one chooses the best alternative (Puccio et al., 2011). These aspects of thinking are in common within other types of thinking. However, the key challenge here is that the problems that require complex thinking are also complex. They are typically ill-defined and new, and depend on multiple variables, which are interconnected. The difficulty lies not only in the number of variables but also in the degree in which they are interconnected.

We can now recognize that a green synthesis, which requires the consideration of the 12 principles of green chemistry, is an example of a complex problem. The interconnection between the 12 principles is dependent also to a degree on the type of synthesis. Thus, complex thinking is well suited for green chemistry applications.

3.2.7   CONCLUSION ABOUT THINKING THAT IS SUITABLE FOR GREEN CHEMISTRY

As we described the various types of thinking, it became clear that we need to depart quickly from the formulaic methods of deductive thinking even though they provide the comfort of known equations and known methods of solution. Although this is the most commonly required type of thinking for the sciences and mathematics, we have shown its limitations. We have presented other types of thinking, from the inductive to complex. These types of thinking are less straightforward and more complicated than deductive thinking. Creative and novel solutions can be generated by ways of multiple thinking approaches. Knowledge of different ways of thinking should allow the students to develop, exercise, and finally contribute creativity and innovation to the field of green chemistry.

3.3   EXAMPLES OF INNOVATIONS IN GREEN CHEMISTRY

We have learned in Section 1.3 that the birth of Green Chemistry itself was an act of innovation, a paradigm shift from pollution control to pollution prevention. In Section 2.2.2, we have covered atom economy, which is an outstanding example of innovation in green chemistry. In this section, we provide a brief description of some other innovations in green chemistry, which we shall expand upon in Chapters 4–7. Many more examples of innovations will be given in Chapters 8–13.

3.3.1 Reactions in Water

As we have already learned, organic solvents may be toxic to humans, harmful to life in general, and damaging to the environment and hazardous in other respects. They are used in massive amounts in chemical industry. Replacing these solvents with benign ones, such as water, or avoiding the use of solvents all together, is a major objective of green chemistry.

We have also learned that a great portion of organic chemistry knowledge is based on the use of organic solvents, which was thus thought to be unavoidable. Green chemistry has overturned this paradigm, as it recognized and exploited the ability of organic compounds to react in water.

In this section, we explain the principles behind organic reactions in water with the highlight on innovation.

The old thinking was that chemists must dissolve organic materials in order for them to react. However, it was known that some organic reactions occur "on water," namely, in a suspension of organic materials in water, in which organic materials are the insoluble components. Such reactions were considered just a curiosity, and there was no theoretical explanation for their occurrence. Ronald Breslow reinvestigated these reactions and explained that they occur due to hydrophobic effects, which are well known in biology (Breslow, 1991). We offer here a metaphoric, simplified, and approximate explanation of these effects (but will cover them in depth and more accurately in Chapters 5 and 13): Chemicals that do not dissolve in water are called hydrophobic, which means water "hating." They try to "escape" water because they "hate" it, and in the process come close together. This proximity facilitates a proper orientation of the molecules and increases the probability of the reaction to occur. Breslow's explanation was important since the reactions that were considered unexplained exceptions now had a theoretical explanation. The papers reporting organic reactions "on water" were finally read, studied, and accepted by chemists who could now run the reactions in water. This realization was of a particular significance for the greening of the industrial reactions and processes. Breslow employed innovative thinking and made connections between the hydrophobic effects and the "on water" reactions that were previously unexplained. He has made such connections against the common belief in organic chemistry that we must dissolve the organic compounds in solvents in order for them to react. This represented an innovation, and another paradigm shift. We shall cover in depth the theoretical foundation of these reactions and will provide specific examples of such reactions in Chapter 4.

3.3.2 Reactions in Superheated Water

There were many papers available in the chemical literature that described organic reactions in superheated water. A very unusual discovery was reported that the changes in the dielectric constant of water which occur at high temperatures and pressures make water behave just like some organic solvents! For example, at 150°C water behaves like dimethylsulfoxide [$(CH_3)_2SO$], at 175°C as *N,N*-dimethylformamide [$(CH_3)_2N–CHO$], at 200°C as acetonitrile (CH_3CN), and at 300°C as acetone [$(CH_3)_2CO$] (Leadbeater and McGowan, 2013). These are all aprotic solvents, whereas water is a protic solvent. The fact that superheated water can behave like acetone, which is as close to the universal organic solvent as we can get, and starts dissolving organic compounds with ease is astonishing! Previously, it was believed that we must use organic solvents because water properties are not suitable for dissolving them. The discovery that water under certain conditions can behave like some organic solvents represented another paradigm shift. Moreover, the pathway of the chemical reaction (the mechanism) in superheated water may become altered, thus resulting in novel reactions. We shall expand on this topic later, in more detail, with the appropriate theoretical background, and with many examples of such reactions, in Chapter 5.

3.3.3 REACTIONS IN WATER IN NANOMICELLES

In 2011, Bruce H. Lipshutz, professor from the University of California, Santa Barbara, received the Presidential Green Chemistry Challenge Award for his project "Towards Ending Our Dependence on Organic Solvents," which resulted in the discovery of the reactions that can be run in the aqueous medium in the nanomicelles (very small micelles that have diameters of 10–100 nm).

We have already covered lots of ground on the topic of organic solvents and their toxicity. Many of these solvents are also volatile and flammable, and are derived from petroleum, a nonsustainable source. We have also introduced water as a benign solvent, and have explained why and how it can substitute for organic solvents (and will discuss more about it in Chapters 4 and 5). However, Bruce H. Lipshutz' innovation provides a conceptually different way, which allows numerous organic reactions to occur in water with excellent results (Lipshutz and Ghoral, 2012). He designed a novel surfactant, named TPGS-750-M, which forms nanomicelles in water. These micelles act as chemical reactors. The structure of this surfactant is shown in Figure 3.2.

Before we can understand the ingenuity of this design, we need to review the basics on terms "surfactant" and "micelle." Surfactants are "amphiphilic" molecules, which means that they possess both a water-soluble component ("hydrophilic" or "water-loving") and a water-insoluble part ("hydrophobic" or "water-hating" part). A schematic representation of an amphiphilic molecule is shown in Figure 3.3.

Students may recall some common uses of surfactants. They include detergents, such as those in dish-washing liquids, which help bring grease into the water phase, and emulsifiers, which are added to oil-and-vinegar salad dressings, to help in mixing the ingredients (see Wikipedia on surfactants).

When surfactants are placed in water, they can organize themselves into the structures called micelles (see Figure 3.4). In micelles, the hydrophobic part of the molecule points inward and the hydrophilic part is oriented outside, toward water (see Wikipedia on micelles).

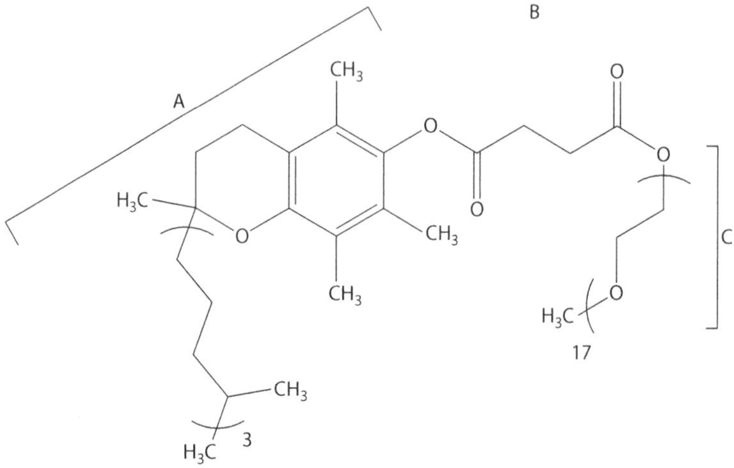

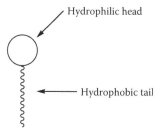

FIGURE 3.2   Chemical structure of TPGS-750-M. A, racemic vitamin E; B, succinic acid; and C, methoxypolyethylene glycol.

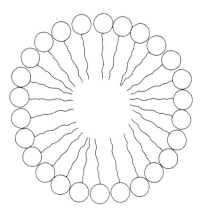

FIGURE 3.3   Schematic representation of an amphiphilic molecule.

FIGURE 3.4   Generalized structure of a micelle, made from an amphiphile. See Figure 3.3 for the representation of the latter.

We now return to the award-winning surfactant, TPGS-750-M, whose structure is shown in Figure 3.2. This surfactant is designed to be green. Its structural components are vitamin E, in a racemic form, which is benign by virtue of being a vitamin; succinic acid, which is also benign because it is an intermediate in cellular respiration; and MPEG-750, a methoxypolyethylene glycol, which is a common degradable hydrophilic compound. Not only are the structural components of TPGS-750-M nontoxic, but they are also inexpensive. Thus, this surfactant's structure is an exemplar of a design that meets the green standard.

When this award-winning surfactant is placed in water, it spontaneously forms very small micelles, which are called nanomicelles, because they have diameters of only 10–100 nm. These micelles are lipophilic ("fat-loving") on the inside and hydrophilic ("water-loving") on the outside. The organic starting materials and catalysts are typically not soluble in water and are thus hydrophobic ("water-hating"), but are at the same time lipophilic. They will go inside the micelles where they will "dissolve" in the lipophilic medium. They are quite concentrated inside the micelles. This results in high reaction rates. The micelles thus act as nanoreactors. The outside part of the micelle interacts favorably with water, because it is hydrophilic, and thus, micelles disperse themselves in water. There are numerous reactions that can be run in these nanomicelles. Because the reaction rates are high at ambient temperature, the energy expenditure (e.g., in the form of heating) is not required. Water does not need to be specially purified. Even seawater can be used. Isolation of products is easy, and the surfactant is easy to recycle. These are additional green benefits of this process.

3.3.4 DESIGN OF HALOGEN-FREE FIREFIGHTING FOAM

We select here the innovation by Solberg Co., from Green Bay, Wisconsin, which received the Presidential Green Chemistry Challenge Award in 2014 for development of halogen-free firefighting foam (Ritter, 2014). We first provide a short background, and then describe the innovation.

Firefighting foams suppress combustion by smothering and cooling fires. Standard critical components of these foams for years have been fluorinated surfactants, with long hydrocarbon chains (eight or more carbons). These surfactants have been shown to be persistent, and are bioaccumulative and toxic. The foam formulators then switched to short-chain (six or less carbons) fluorosurfactants, which were shown to be less toxic and less bioaccumulative, but were still persistent. A drawback of the short-chain design was that one had to use almost 40% more of the short-chain fluorosurfactants compared to the long-chain ones to achieve the required firefighting performance. This was the state of the art in this field, until the Solberg Co. made a major innovation, consisting of a replacement of the fluorosurfactants with a blend of nonfluorinated, bio-based surfactants and complex carbohydrates. The new foam, RE-HEALING™, performs equally well or better than the fluorinated foams. It has more capacity to absorb heat, due to the presence of complex carbohydrates.

The bio-based surfactants in the new foam are derived from plant sugars and oils, rather than from petroleum, and are thus renewable. These surfactants are already in use in household and personal care products, such as soaps and tooth-paste. The foam also contains complex carbohydrates such as sugar molasses and a proprietary polysaccharide, among some other ingredients. The components of RE-HEALING are benign. This award-winning foam degrades completely in the environment after 6 weeks. This example dramatically illustrates the innovation principle in which lateral thinking was employed. Rather than digging the vertical hole deeper and deeper, namely, tweaking the size of the carbon chain of fluorosurfactants, a new hole was dug, which used green bio-based materials instead (see Section 3.2.5 on lateral thinking for the metaphor on digging holes).

3.3.5 Sources of Examples of Innovation in Green Chemistry

There are many examples of innovation in green chemistry, and we strongly encourage students and other readers to research as many as possible. They are informative, fascinating, and inspiring. We recommend as an excellent source the U.S. Environmental Protection Agency (EPA) website, which provides information about the Presidential Green Chemistry Challenge Award (see the EPA website: www2.epa.gov/green-chemistry). This award program was created in 1995 to promote the use of green chemistry for pollution prevention. The awards are given each year, one of each in the following categories: Academia, Small Business, Greener Synthetic Pathways, Greener Reaction Conditions, and Designing Greener Chemicals. The winners are individuals, groups, and organizations, which designed, developed, and implemented green chemistry technology that is innovative and economically feasible, and contributes to pollution prevention goals (see, e.g., the EPA website: www2.epa.gov/green-chemistry; Hill et al., 2010).

REVIEW QUESTIONS

Individual assignments within class projects are given below. For each innovation, discuss which type of thinking was used (in your opinion).

3.1 Find one example of innovation for a small-business Presidential Green Chemistry Challenge Award. Research the innovation further and report your findings to your colleagues.

3.2 As you research the innovations for the Presidential Green Chemistry Challenge Award, you will notice the option to select an industry that has received the award. Describe a winning technology in the agriculture and agrochemicals field, for the topic of pesticides.

3.3 In 2004, Bristol-Myers Squibb Company received an award for the synthesis of Paclitaxel, an active ingredient in Taxol®, a drug used to treat ovarian and breast cancers, by plant cell fermentation. You will find a summary of the invention on the EPA website. Research this innovation and present the results to the class.

3.4 In 2013, Cargill Inc., (Minnetonka, MN), received a Presidential Green Chemistry Challenge Award for the dielectric coolant made from vegetable oil, which replaced mineral oil, polychlorinated biphenyls (PCBs), or other halogenated compounds in electrical power transformers. Research this winning technology, starting by a green chemistry analysis. What was wrong with PCBs or other halogenated compounds in electrical power transformers from the green chemistry's point of view?

ANSWERS TO REVIEW QUESTIONS

Answers are provided on the EPA website, in the summary of each award. As for the type of thinking that was involved in the innovation, class discussion will help reveal multiple types of thinking.

REFERENCES

About innovation and invention (accessed on September 11, 2014), http://en.wikipedia.org/wiki/Innovation; http://en.wikipedia.org/wiki/Invention; www.merriam-webster.com/dictionary/innovation; www.businessdictionary.com/definition/innovation.html; www.conferenceboard.ca/cbi/innovation.aspx

About surfactants and micelles (accessed on October 23, 2014), https://en.wikipedia.org/wiki/Surfactant; https://en.wikipedia.org/wiki/Micelle

About the Presidential Green Challenge Awards (accessed on October 22, 2014), www2.epa.gov/green-chemistry; open this site and explore the links to the awards sections.

About various thinking styles (assessed on September 24, 2014), http://en.wikipedia.org/wiki/Lateral_thinking; http://en.wikipedia.org/wiki/Critical_thinking; http://en.wikipedia.org/wiki/Integrative_thinking; http://en.wikipedia.org/wiki/Systems_thinking

Breslow, R. (1991). Hydrophobic effects on simple organic reaction in water, *Acc. Chem. Res.*, 24, 159–164.

de Bono, E. (1990). *Lateral Thinking: Creativity Step by Step*, Harper Perennial, Harper & Row Publishers, New York, pp. 11–14, 42–43, 106, 123.

de Bono, E. (1993). *Serious Creativity: Using the Power of Lateral Thinking to Create New Ideas*, Harper Business, Harper Collins Publishers, New York, pp. 52–56.

Hill, J. W., McCreary, T. W., and Kolb, D. K. (2010). *Chemistry for Changing Times*, 12th edn., Prentice Hall, New York, pp. 31, 55, 115, 151, 211, 250, 284, 440, 464, 530.

Kahn, C. H. (1995 reprint of 1979 edition). *The Art and Thought of Heraclitus*, Cambridge University Press, Cambridge, p. 69.

Lawhead, W. F. (2001). *The Voyage of Discovery: A Historical Introduction to Philosophy*, 2nd edn., Wadsworth/Thomson Learning, Belmont, CA, see especially p. 215. See also the original work of Francis Bacon, *The New Organon* (various editions exist), which is surprisingly readable, and it shows nicely the application of the inductive method to scientific thinking.

Leadbeater, N. and McGowan, C. B. (2013). *Laboratory Experiments Using Microwave Heating*, CRC Press, Boca Raton, FL, p. 6.

Lipshutz, B. H. and Ghoral, S. (2012). "Designer"-surfactant-enabled cross-couplings in water at room temperature, *Aldrichimica Acta*, 45(1), 3–16.

Nickerson, R. S., Perkins, D. N., and Smith, E. E. (1985). *The Teaching of Thinking*, Lawrence Erlbaum Associates, Hillsdale, NJ; see especially Chapter 5, Errors and biases in reasoning, pp. 111–142.

Osterman, M. Reio, Jr., T. G, and Thirunarayannan, M. O. (2013). Digital literacy: A demand for nonlinear thinking styles, in Plakhotnik, M. S. and Nielsen, S. M., eds., *Proceeding of the 12th Annual South Florida Educational Research Conference*, Florida International University, Miami, FL, pp. 149–154. http://education.fiu.edu/research_conference, accessed on September 25, 2014.

Padgett, J. F. and Powell, W. W. (2012). *The Emergence of Organizations and Markets*, Princeton University Press, Princeton, NJ, pp. 1–15.

Puccio, G. J., Mance, M., and Murdock, M. C. (2011). *Creative Leadership: Skills That Drive Change*, 2nd edn., Sage Publications, Thousand Oaks, CA, pp. 55–56.

Reif, F. (2010). *Applying Cognitive Science to Education: Thinking and Learning in Scientific and Other Complex Domains*, MIT Press, Cambridge, MA.

Ritter, S. K. (2014). "2014 Green Chemistry Awards," *Chemical and Engineering News*, October 20, pp. 32–34.

4 Green Organic Reactions "on Water"

... the major real reason to pursue water as a solvent is that the hydrophobic effect leads to such remarkable new chemistry not otherwise achievable.

Ronald Breslow (2010)

LEARNING OBJECTIVES

The learning objectives for this chapter are as follows.

Learning Objectives	Section Numbers
Learn about different types of reactions in the aqueous medium	4.1
Learn about specific examples of on-water reactions	4.2
Diels–Alder reaction	4.2.1
Passerini reaction	4.2.2
A sampler of other "on-water" reactions	4.2.3

4.1 TYPES OF REACTIONS IN AQUEOUS MEDIUM AND ASSOCIATED NOMENCLATURE

We start with the simplest and most familiar concept of organic reactions in water. Some categories of organic compounds are soluble in water. Such compounds include small molecules, typically five carbon atoms or less, which have hydrophilic groups present, for example, hydroxyl (–OH), carboxyl (–COOH), or amino (–NH$_2$). Charged organic compounds, such as salts of carboxylic acids (carboxylates; –COO$^-$), phenols (phenolates; PhO$^-$), and amines (amine salts; –NH$_3{}^+$), are typically also soluble. This is the case providing that the hydrophobic portion of these salts is not disproportionally large. Familiar examples of water-soluble organic compounds include simple sugars such as glucose, amino acids such as glycine, and alcohols such as ethanol or methanol. Water-soluble compounds can react in water. Such reactions are classical in-water reactions.

This simple concept of organic reactions in water has been the accepted paradigm until relatively recently. Because the vast majority of organic compounds are not water soluble, this paradigm implied that such compounds cannot react in water. This view changed in 1980, when it was demonstrated that water-insoluble organic compounds can react in water, but that water can also accelerate such reactions

(Breslow and Rideout, 1980). Much more work followed, which showed that this phenomenon is widespread (Klijn and Engberts, 2005; Narayan et al., 2005; Chandra and Fokin, 2009). These findings brought a whole new meaning to the concept of organic reactions in water.

A new nomenclature that would make a distinction between in-water reactions of soluble versus insoluble reactants was needed. The new nomenclature introduces terms "on water," "in water," "water-promoted," and "in the presence of water," among others (Lubineau et al., 1994; Shapiro and Vigalok, 2008; Chandra and Fokin, 2009). These terms are often used interchangeably, although they refer to reactions that may occur under quite different conditions. The nomenclature that is used most often is as follows:

The "in-water" reactions describe the cases in which the reactants are water soluble, but also the cases in which the reaction occurs in micelles that are dispersed in water (see Section 3.3.3 for the award-winning micelle design by Lipshutz).

The "on-water" reactions are those in which the insoluble reactants are stirred in aqueous emulsions or suspensions. Thus, they occur in heterogeneous systems (Klijn and Engberts, 2005; Narayan et al., 2005; Chandra and Fokin, 2009).

"Water-promoted" organic reactions are those that show an outcome that is superior when the reactions are performed in water compared to traditional organic solvents.

"In the presence of water" can be used broadly, but also more specifically, for the cases in which water is added to traditional organic solvents.

In this chapter, the focus is on the on-water reactions. However, some such reactions may also be covered under other categories, such as under water-promoted.

In Section 3.3.1, we have addressed the innovative aspect of organic reactions in water that occur between water-insoluble organic compounds. Now we can term such reactions properly as "on water." In the same section, we have described in simple terms the hydrophobic effects, which enable reactions of water-insoluble reactants to occur in water. Now we introduce a common nomenclature for the hydrophobic effects and the associated phenomena, which will facilitate reading of the rest of this chapter. We use Breslow's work as the major source for the nomenclature (e.g., Breslow, 1980, 1991, 2007).

Hydrophobic effect: The tendency of nonpolar compounds or their segments to cluster in water, to diminish the interface with water.

Hydrophobic acceleration: The acceleration of the reaction rate due to the hydrophobic effect.

Prohydrophobic additives: Additives that increase hydrophobic effects. They decrease the solubility of nonpolar reactants in water.

Antihydrophobic additives: Additives that decrease hydrophobic effects. They increase the solubility of nonpolar reactants in water.

4.2 EXAMPLES OF ON-WATER REACTIONS

In this section, we choose two examples of the on-water reactions, which are well known, are well understood, and have extensive applications. They are the Diels–Alder and Passerini reactions. We also offer a sampler of other on-water reactions, to illustrate the breadth and scope of this class of reactions.

4.2.1 DIELS–ALDER REACTION

The Diels–Alder reaction is one of the more important carbon–carbon bond-forming reaction classes. This reaction occurs between a diene and a dienophile. These two starting materials add to each other to give a single product, which is referred to as a Diels–Alder adduct. We have shown an example of a Diels–Alder reaction in Figure 2.7, in conjunction with calculation of atom economy. For this reaction, the atom economy is 100%. From Figure 2.7, which shows the reaction between cyclopentadiene (the "diene" component) and maleic anhydride (the "dienophile" component), we see another key feature of this reaction. It is the capability to form a six-membered ring, a cyclohexene. The double bond in the latter can be used as a "handle" for additional reactions, if desired. Diels–Alder reaction is stereospecific, because it gives a preferred isomer, the so-called *endo* isomer. This stereospecificity is the result of the required electronic orbital overlap between the reactant molecules. This is shown in Figure 4.1, for the same reaction example as depicted in Figure 2.7.

This reaction is fully described in any beginning organic chemistry textbook (e.g., Solomons and Fryhle, 2011).

Traditionally, the Diels–Alder reaction has been performed in organic solvents. However, in many cases, it may occur in water too. The water-insoluble hydrophobic reactants, when placed in water, are driven toward each other to avoid water. In this process, they experience hydrophobic interactions. These are well known in biology, especially for the interaction of peptides. Such hydrophobic interactions increase the probability of a close interaction between reactants. This process is sometimes referred to as the increase in the effective concentration due to hydrophobic "packing." The close proximity of the reactants also facilitates a proper alignment of their

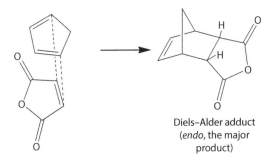

Diels–Alder adduct
(*endo*, the major product)

FIGURE 4.1 Diels–Alder reaction between cyclopentadiene and maleic anhydride. Orbital overlap that leads to the major product (*endo*) is shown.

molecular orbitals, which increases both the reaction rate and stereospecificity. When the reaction is performed in organic solvents, the reactants dissolve in the solvent and become diluted. This dilution decreases the probability of the reactants to encounter each other, which is reflected in the decreased reaction rate. The *endo*-specificity of the reaction is also adversely affected in this case.

The operation of the hydrophobic effect in Diels–Alder on-water reactions was not proposed just by analogy with the well-known cases of this effect in protein chemistry, or other examples from biology. There are multiple lines of experimental evidence which support the operation of the hydrophobic effect in Diels–Alder reactions. Figure 4.2 shows the equation for the Diels–Alder reaction, which was studied extensively by Breslow and coworkers as a model reaction for a demonstration of hydrophobic effects. This reaction involves anthracene-9-methanol as the diene and *N*-ethylmaleimide as the dienophile.

The operation of hydrophobic effects is supported by modulation in reaction properties through the additions of various compounds (so-called "additives"). The latter may be salts which are known for decreasing the water solubility of organic

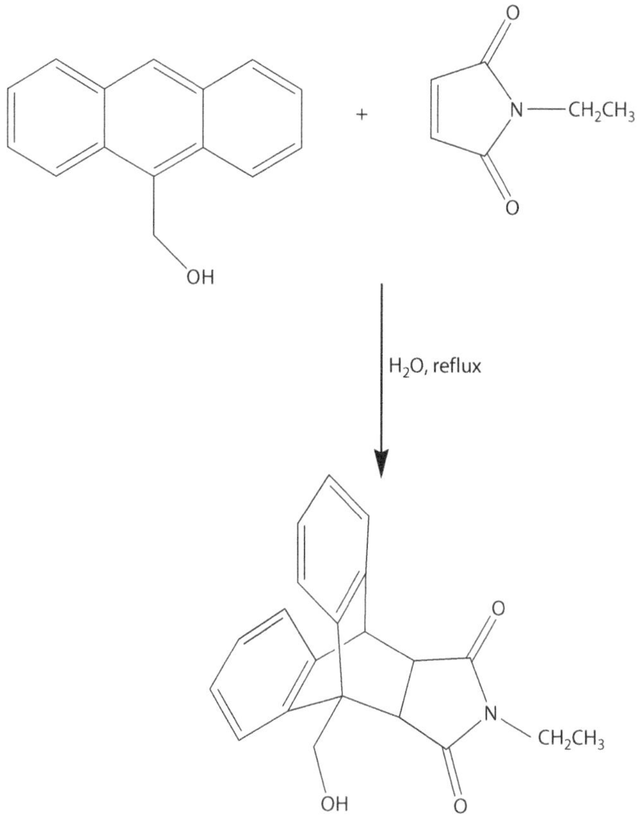

FIGURE 4.2 Diels–Alder on-water reaction between anthracene-9-methanol and *N*-ethylmaleimide.

compounds. These are called prohydrophobic agents. Examples include NaCl or LiCl. These agents increase hydrophobicity, and thus also the reaction rate. Addition of antihydrophobic agents, such as guanidinium chloride, has the opposite effect. For example, the reaction rate of Diels–Alder reaction shown in Figure 4.2 is 2.5 times faster in LiCl and 3 times slower in guanidinium choloride, compared to pure water (Breslow, 1991).

Table 4.1 shows examples of prohydrophobic and antihydrophobic additives. From Table 4.1, we can see that smaller ions, such as in LiCl or NaCl, increase hydrophobicity, whereas larger ions decrease it, such as in the cases of I^- (within LiI), or ClO_4^- ions.

In 1992, Rizzo investigated the relationship between the rate of the reaction shown in Figure 4.2 in aqueous salt solutions and the size of the anion within the salt. He plotted the reaction rate versus crystallographic ionic radius and has found that as the radius of the anion increases, the reaction rate decreases in a linear fashion. These results are significant because they show the quantitative regularity of the hydrophobic effect as observed in the so-called hydrophobic acceleration of Diels–Alder reactions. This opens the door for utilization of salt effects as a diagnostic tool for the operation of hydrophobic effects.

A question arises if the action of additives is applicable only to the Diels–Alder reaction, or it is more general. The answer is affirmative. Breslow (1991) gave another example, a reaction that students are likely familiar with: the benzoin condensation. We shall discuss this reaction in Chapter 13, in context of green catalysis.

The Diels–Alder reaction from Figure 4.2 became a standard green experiment in the undergraduate organic laboratory (see Chapter 13).

There are more benefits of running the Diels–Alder reaction on water other than the increase of the reaction rate, although the latter may be more than 10,000 times! (Otto and Engberts, 2000). Water also enhances the selectivity of the reaction, notably the formation of the already preferred *endo* product. This enhanced selectivity may be attributed to the difference in stability of the transition states for the *endo* as opposed to the *exo* isomer in water. Hydrophobic effects are assumed to favor stabilization of the *endo* transition state, because it is more compact. In contrast, the transition state for the *exo* product is more extended. A concise explanation is that "hydrophobic packing" favors the *endo* product.

The application of the *endo* preference in water has a green chemistry application. When we get two products, *endo* and *exo*, where *endo* is the usual desired product, the atom economy of the *endo* product increases in water.

TABLE 4.1
Examples of Prohydrophobic and Antihydrophobic Additives

Prohydrophobic Additives	Antihydrophobic Additives
LiCl, NaCl, NaBr, KCl	Guanidinium chloride, $(NH_2)_2C=NH_2^+Cl^-$
$MgCl_2$, $CaCl_2$, Na_2SO_4	Urea, $(NH_2)_2C=O$, ethanol
Glucose, saccharose, sucrose	LiI, perchlorate ion, ClO_4^-

Breslow made a successful and inspired generalization of the action of the hydrophobic effect from known cases in biology to that of small organic molecules. For example, he cites hydrophobic effects seen in proteins, nucleic acids, and the binding of antigens to antibodies. Additional biological examples are the formation of the micelles and bilayers. This generalization is the result of complex thinking in which connections were made between the phenomena from different disciplines. It resulted in a major innovation for the green chemistry field.

An important question arises at this point: Is the Diels–Alder on-water reaction the only one that exhibits an aqueous rate enhancement or are there other similar reactions? The answer is that the Diels–Alder reaction is not unique in this respect, as many different types of reactions exhibit the same effect. Examples of such reactions will be presented in the subsequent sections of this chapter.

4.2.2 PASSERINI REACTION

The Passerini reaction was chosen for three major reasons. First, this reaction is pedagogically well developed into a green on-water undergraduate laboratory experiment (Hooper and De Boef, 2009; Willamson and Masters, 2011; see also Chapter 13). Second, the Passerini reaction is representative of a broad class of so-called multicomponent reactions that can be performed on water and that exhibit other green features. The on-water behavior of this reaction has been studied extensively (e.g., Pirrung and Sarma, 2004; Sela and Vigalok, 2012). The third reason for studying this reaction is for its use in applied chemistry, for example, in the pharmaceutical industry (Dömling and Ugi, 2000; Dömling, 2006) and in polymer science (Kreye et al., 2011).

An example of the Passerini reaction that became a green undergraduate laboratory experiment (see Chapter 13) is shown in Figure 4.3.

The Passerini reaction is an old reaction. It was discovered in 1921 by Mario Passerini, an Italian chemist, after whom it was named. The fact that this reaction is "named" (just like the Diels–Alder reaction) indicates that it is frequently used, and thus chemically useful. A lengthy review of this reaction (140 pages) was published in the "Organic Reactions" book series (Banfi and Riva, 2005), which is often the ultimate source of information about important organic reactions. The beginning of that review addresses the mechanism of the reaction and offers various answers to the intriguing question: How do the three reactants get together? The probability of all three combining at once is rather low. Most of the discussion is relevant to the "classical" Passerini reaction, which was not performed in water. Instead, the reaction was run in nonpolar or low polarity solvent, such as dichloromethane, diethyl ether, and ethyl acetate, among others. Only a limited number of reactions were run in more polar, but aprotic solvents, such as dimethylformamide or dimethylsulfoxide. Importantly, alcohols such as methanol and ethanol, which are polar protic solvents, are not well suited for the Passerini reaction. Only toward the end of the review is the discovery of the Passerini reaction in water addressed. This discovery was made in 2004; thus, quite recent for the review was published in 2005. This tells us that the green chemistry application of the Passerini reaction in water came after more than eight decades after its original publication. This green chemistry insight

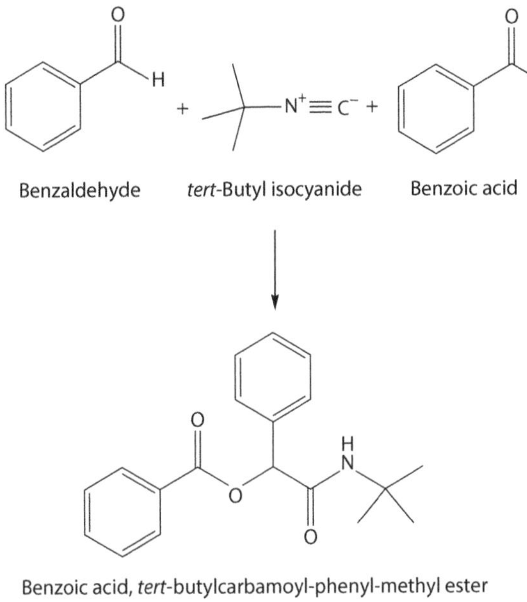

Benzaldehyde   *tert*-Butyl isocyanide   Benzoic acid

Benzoic acid, *tert*-butylcarbamoyl-phenyl-methyl ester

FIGURE 4.3 Passerini reaction between benzaldehyde, *tert*-butyl isocyanide, and benzoic acid.

was rather spectacular, because it showed that this reaction not only occurs in the aqueous medium (on water), but that it shows a remarkable reaction rate acceleration compared to dichloromethane as a solvent (Pirrung and Sarma, 2004). This acceleration does not appear to be related to the polarity of water, because the reaction does not go well in alcohols, which are polar protic solvents similar to water. Instead, the acceleration is due to the hydrophobic effects. We can now answer the questions we have asked earlier: How do the starting materials of the Passerini reaction get together and how is the low probability of such an event overcome? The answer to both questions is the aforementioned "hydrophobic effect" or "hydrophobic packing." This is analogous to the Diels–Alder reaction on water, in which the dienes and dienophiles are packed together by the hydrophobic effect. As for the details of the mechanism of the Passerini reaction, much can be learned from the recent study by Pirrung and Sarma (2004). Notably, the Diels–Alder and Passerini reactions are linked by their common hydrophobic effect. However, these authors also note the effect of the high cohesive energy density of water and the influence of temperature along with other factors as contributors to the observed hydrophobic effect.

Pirrung and Sarma used as a prototype the Passerini reaction shown in Figure 4.4. They have found that the reaction performed in dichloromethane at 25°C for 18 h gives 50% conversion and an eventual 45% yield. In contrast, the reaction run in water at the same temperature for 3.5 h gives 100% conversion and a 95% yield. The authors have also measured the reaction rates. The reaction in water is approximately 18 times faster than that in dichloromethane. These combined results show additional advantages of running the reaction in water, above and beyond eliminating

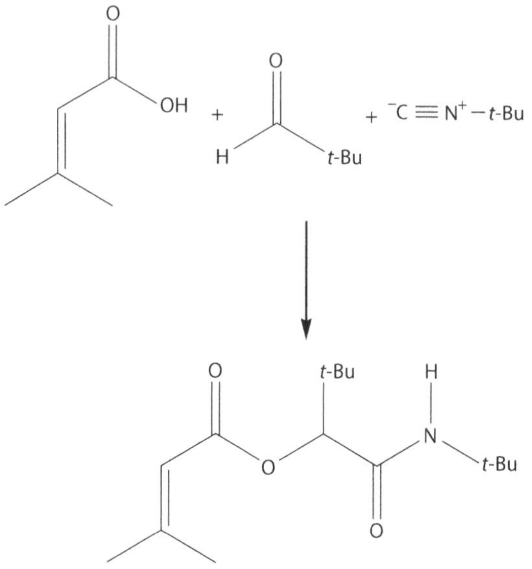

FIGURE 4.4 Passerini reaction used as a prototype by Pirrung and Sarma. (From Pirrung, M. C. and Sarma, K. D., Multicomponent reactions are accelerated in water, *J. Am. Chem. Soc.*, 126, 444–445, 2004. With permission.)

the use of toxic dichloromethane. Next, the authors have investigated the reaction in water with different additives and under different temperatures. In all cases, the conversion was 100% and the yields were high (91%–95%). Again, the reaction rates were measured. The addition of prohydrophobic additives, such as LiCl and glucose, substantially increased the reaction rates. In the case of LiCl, the factor was 16 or more, and for glucose 7 or more, depending on the concentration of the additives. This reaction increase supports the operation of hydrophobic effects in this reaction.

In addition, Pirrung and Sarma have found a fascinating temperature effect on the Passerini reaction in water. When the temperature was increased from 25°C to 50°C, the reaction rate decreased for 44%. When the temperature was lowered from 25°C to 4°C, the reaction rate increased by 11%.

This type of temperature effects may not be familiar to students, whose laboratory experience may be to the contrary. Most common organic reactions exhibit a rate increase upon heating and a rate decrease upon cooling. The temperature effects that are observed in the Passerini reaction are called "inverse temperature dependence." They provide additional insights into the nature and the mechanism of this reaction, and point to the possible importance of the cohesive energy density of water, among other factors (Pirrung and Sarma, 2004). These topics are suitable for more advanced students and are addressed in the References section.

The inverse temperature dependence of the Passerini reaction has mostly favorable consequences for green chemistry. Performing this reaction at room temperature gives a good outcome, heating makes the outcome worse, and cooling improves it only modestly. The energy cost required to bring the temperature down from 25°C to 4°C needs to be weighed against the benefit of an 11% increase in the reaction rate. For most routine experiments, such cooling would not pay.

Let us now return to the Passerini reaction from Figure 4.3, which is one of the popular undergraduate laboratory experiments (see Chapter 13), to point out another green advantage of this reaction. When this reaction is performed on water, the reaction product is insoluble in water. All that is needed to isolate the product is a simple vacuum filtration of the solid product. Thus, no solvents will be needed for the isolation of the product. Because the reaction has an atom economy of 100%, and an almost quantitative yield, you may not even need to purify the crude product.

4.2.3 A Sampler of Other "On-Water" Reactions

Numerous on-water reactions have been reported (Narayan et al., 2005; Chandra and Fokin, 2009; Liu and Wang, 2010). They include various cycloadditions, Claisen rearrangements, nucleophilic ring opening of epoxides, oxidations, reductions, brominations, and various nucleophilic substitutions, among others. We present here chemical equations for selected common, classical reactions in which the on-water effect will transform them into green chemistry reactions. Students are likely familiar with these reactions, because they are found in any basic organic chemistry textbook (e.g., Solomons and Fryhle, 2011).

Figure 4.5 shows an example of Claisen rearrangement, which was performed on water at room temperature. It gave a 100% yield. The same reaction in an organic solvent or neat (without solvent) gave a yield of ~70% or lower (Narayan et al., 2005). Figure 4.6 depicts a nucleophilic opening of an epoxide. This reaction also occurred much faster on water than in organic solvents or under neat conditions (no solvent) (Narayan et al., 2005). Figure 4.7 presents the air oxidation of an aldehyde to a carboxylic acid (Shapiro and Vigalok, 2008; Liu and Wang, 2010; Sela and Vigalok, 2012). The aldehydes that were used are typically hydrophobic aliphatic aldehydes, which can be linear, branched, or cyclic, or aromatic aldehydes.

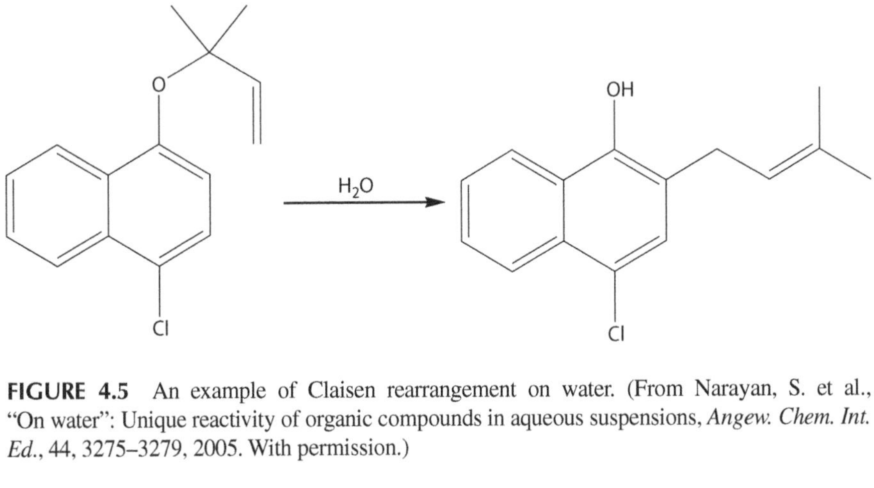

FIGURE 4.5 An example of Claisen rearrangement on water. (From Narayan, S. et al., "On water": Unique reactivity of organic compounds in aqueous suspensions, *Angew. Chem. Int. Ed.*, 44, 3275–3279, 2005. With permission.)

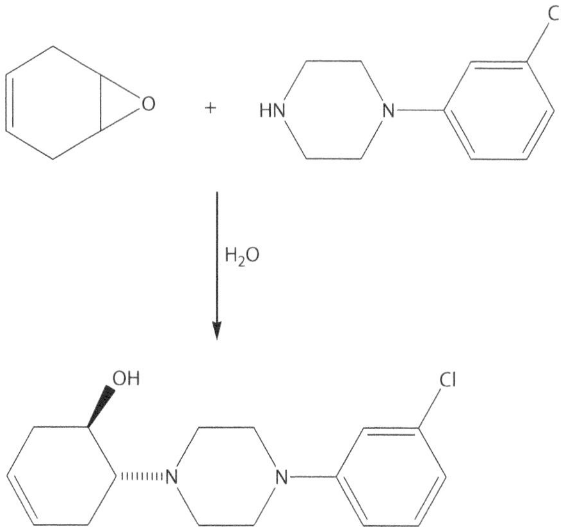

FIGURE 4.6   An example of a ring opening of an epoxide by a nucleophile. (From Narayan, S. et al., "On water": Unique reactivity of organic compounds in aqueous suspensions, *Angew. Chem. Int. Ed.*, 44, 3275–3279, 2005. With permission.)

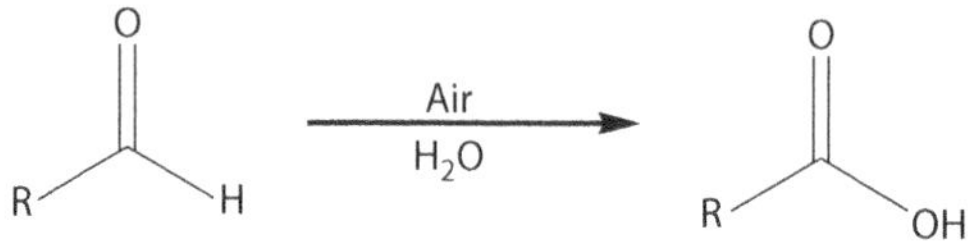

R = Cyclohexyl, *n*-butyl, *n*-heptyl, 2-ethylhexyl, and Ph

FIGURE 4.7   Air oxidation of aldehydes on water. (Liu, L. and Wang, D.: "On water" for green chemistry. *Handbook of Green Chemistry*. 2010. 207–228. Li, C.-J., ed. Copyright Wiley-VCH Verlag GmbH & Co. KGaA; Shapiro, N. and Vigalok, A., Highly efficient organic reaction "on water," "in water," and both, *Angew. Chem.*, 120, 2891–2894, 2008. With permission.)

REVIEW QUESTIONS

4.1 With regard to the Passerini reaction, which principles of green chemistry are involved in the experimental observations of the following:
   a. 100% atom economy
   b. Good results at room temperature
   c. The isolation of the product by a simple filtration
   d. The crude product is pure enough so that it does not require recrystallization.
4.2 Consider again the Passerini reaction from Figure 4.4. This reaction occurs at room temperature. You wish to speed up the reaction and thus decide to heat the reaction mixture to 50°C. Is this a good idea?
4.3 Consider Diels–Alder reaction on water from Figure 4.2.
   a. What is its atom economy?
   b. Which additives could you use to speed up this reaction?
   c. Among the possible additives, which one would be the most green?

4.4 Consider the Claisen rearrangement shown in Figure 4.5. What is its atom economy?

4.5 Consider the reaction from Figure 4.6, which represents a ring opening of an epoxide with a nucleophile.
   a. What is the nucleophilic center in this reaction?
   b. What is its atom economy?

4.6 Consider the examples of reactions from this chapter. Which types of reactions show atom economy of 100%?

4.7 In the air oxidation of aldehydes shown in Figure 4.7, pure oxygen rather than air (which is ~21% oxygen) may be used. Compare these two options based on the green chemistry principles.

ANSWERS TO REVIEW QUESTIONS

4.1 a. Principle 2
    b. Principle 6
    c. Principle 3
    d. Principle 3

4.2 No, because this reaction would occur slower at the higher temperature.

4.3 a. 100%
    b. See Table 4.1 for the choices of prohydrophobic additives
    c. NaCl

4.4 100%

4.5 a. The N-lone electron pair from N–H
    b. 100%

4.6 They are (cyclo)additions, multicomponent reactions, and rearrangements.

4.7 One needs to balance the cost of pure oxygen versus free oxygen from the air. Pure oxygen also requires special handling due to its fire and explosion hazards.

REFERENCES

Banfi, L. and Riva, R. (2005). The Passerini reaction, in *Organic Reactions*, Vol. 65, Overman, L. E. et al., eds., Wiley, New York, pp. 1–140.

Breslow, R. (1991). Hydrophobic effects on simple organic reactions in water, *Acc. Chem. Res.*, 24, 159–164.

Breslow, R. (2006). The hydrophobic effect in reaction mechanism studies and in catalysis by artificial enzymes, *J. Phys. Org. Chem.*, 19, 813–822.

Breslow, R. (2007). A fifty-year perspective on chemistry in water, Chapter 1, in *Organic Reactions in Water: Principles, Strategies and Applications*, Lindström, U. M., ed., Blackwell Publishing, Oxford, pp. 1–28.

Breslow, R. (2010). The principles of and reasons for using water as a solvent for green chemistry, in *Handbook of Green Chemistry*, Vol. 5: *Reactions in Water*, Li, C. J., ed., Wiley-VCH, Weinheim, Germany, pp. 1–29; the citation is from p. 25.

Breslow, R. and Rideout, D. (1980). Hydrophobic acceleration in Diels-Alder reactions, *J. Am. Chem. Soc.*, 102, 7816–7817.

Chandra, A. and Fokin, V. V. (2009). Organic synthesis "on water," *Chem. Rev.*, 109, 725–748.

Dömling, A. (2006). Recent developments in isocyanide based multicomponent reactions in applied chemistry, *Chem. Rev.*, 106, 17–89.

Dömling, A. and Ugi, I. (2000). Multicomponent reactions with isocyanides, *Angew. Chem. Int. Ed.*, 39, 3168–3210.

Engberts, J. B. F. N. (2007). Structure and properties of water, Chapter 2, in *Organic Reactions in Water: Principles, Strategies and Applications*, Lindström, U. M., ed., Blackwell Publishing, Oxford, pp. 29–59.

Hooper, M. H. and De Boef, B. (2009). A green multicomponent reaction for the organic chemistry laboratory, *J. Chem. Ed.*, 86, 1077–1079.

Klijn, J. E. and Engberts, J. B. F. N. (2005). Fast reactions "on water," *Nature*, 435, 746–747.

Kreye, O., Tóth, T., and Meier, M. A. R. (2011). Introducing multicomponent reactions to polymer science: Passerini reaction of renewable monomers, *J. Amer. Chem. Soc.*, 133, 1790–1792.

Li, C.-J. and Chen, L. (2005). Organic chemistry in water, *Chem. Soc. Rev.*, 35, 68–82.

Liu, L. and Wang, D. (2010). "On water" for green chemistry, in *Handbook of Green Chemistry*, Li, C.-J., ed., Wiley-VCH, Weinheim, Germany, pp. 207–228.

Lubineau, A., Augé, J., and Queneau, Y. (1994). Water-promoted organic reactions, *Synthesis*, 8, 741–760.

Narayan, S., Fokin, V. V., and Sharpless, K. B. (2007). Chemistry "On water"—Organic synthesis in aqueous suspension, in *Organic Reactions in Water: Principles, Strategies and Applications*, Lindström, U. M., ed., Blackwell Publishing, Oxford, pp. 350–365.

Narayan, S., Muldoon, J., Finn, M. G., Fokin, V. V., Kolb, H. C., and Sharpless, K. B. (2005). "On water": Unique reactivity of organic compounds in aqueous suspensions, *Angew. Chem. Int. Ed.*, 44, 3275–3279.

Otto, S. and Engberts, J. B. F. N. (2000). Diels-Alder reactions in water, *Pure Appl. Chem.*, 72, 1365–1372.

Pirrung, M. C. and Sarma, K. D. (2004). Multicomponent reactions are accelerated in water, *J. Am. Chem. Soc.*, 126, 444–445.

Rizzo, C. J. (1992). Salt effects on a hydrophobically accelerated Diels-Alder reaction follow the Hofmeister series, *J. Org. Chem.*, 57, 6382–6384.

Sela, T. and Vigalok, A. (2012). Salt-controlled selectivity in "on water" and "in water" Passerini-type multicomponent reactions, *Adv. Synth. Catal.*, 354, 2407–2411.

Shapiro, N. and Vigalok, A. (2008). Highly efficient organic reaction "on water," "in water," and both, *Angew. Chem.*, 120, 2891–2894.

Solomons, T. W. G. and Fryhle, C. B. (2011). *Organic Chemistry*, 10th edn., John Wiley & Sons, New York.

Williamson, K. L. and Masters, K. M. (2011). *Macroscale and Microscale Organic Experiments*, 6th edn., Brooks/Cole, Belmont, CA, pp. 699–701.

5 Green Organic Reactions in Superheated Water

Water near its critical point possesses properties very different from those of ambient liquid water.

Phillip E. Savage (1999)

Supercritical water is a special and very extraordinary solvent, as it consists of polar molecules but behaves like a nonpolar solvent.

A. Kruse and E. Dinjus (2012a)

LEARNING OBJECTIVES

The learning objectives for this chapter are as follows.

Learning Objectives	Section Numbers
Understand the fundamental properties of water and the basic features of the phase	5.1
diagram for water, with a special focus on the regions for superheated water.	5.1.1
Become familiar with the terms related to the phase diagrams and superheated water.	5.1.2
Become familiar with the special properties of superheated water, which make it	5.2
a green reaction medium. Recognize and appreciate the innovative aspects of	5.2.1
performing organic reactions in superheated water.	5.2.2
Become familiar with types of organic reactions, which can be performed in	5.3
superheated water.	

5.1 ABOUT WATER: A BRIEF REVIEW

To understand how water behaves when superheated, we need to review some relevant properties of water and its phase diagram, which will be accomplished in Sections 5.1.1 and 5.1.2, respectively.

5.1.1 FUNDAMENTAL PROPERTIES OF WATER

We first show a familiar Lewis structure of water, which depicts the oxygen lone electron pairs (Figure 5.1). These are responsible for the ability of water to act as a hydrogen (proton) acceptor. Due to a large electronegativity difference between oxygen and hydrogen, which makes the hydrogens partially positive, water also acts as a hydrogen donor. This enables water to make hydrogen-bonded networks between

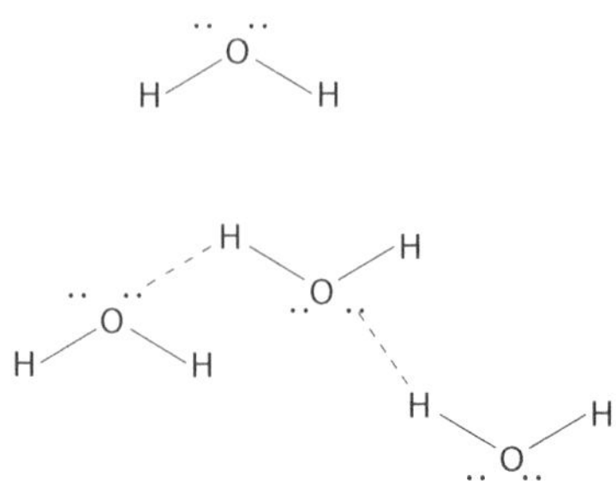

FIGURE 5.1 Lewis structure of water (top) and a network of hydrogen-bonded water molecules (bottom).

its molecules (see Figure 5.1). These networks are responsible for a relatively high boiling point of water, among other properties.

Water is highly polar which is reflected in its high dielectric constant (78.4 at 25°C). Water dissociates to proton (H⁺, which hydrates to H_3O^+) and hydroxide (OH⁻) ions. The ionic product of water ($K_w = [H^+][OH^-]$) is very small. At room temperature, it is 10^{-14}. This means that the concentrations of both protons and hydroxide ions are very small (10^{-7} M). Water is neutral, because the concentrations of both acidic protons and basic hydroxide ions are the same.

As we shall see in the later sections, all these properties of water will be dramatically changed when water becomes superheated.

5.1.2 Phase Diagram of Water and Associated Terms and Definitions

The phase diagram of water is shown in Figure 5.2, and the associated terms are defined in Table 5.1. We shall introduce and explain these terms in the text. Careful study of Table 5.1 will facilitate the understanding of the diagrams.

Each solid line in the phase diagram (Figure 5.2) represents the phase boundary. An equilibrium condition exists between the two phases on either side of the line. Interestingly, the solid–liquid equilibrium line for water has a negative slope, which is unusual. For most other substances, for example, CO_2, the slope for this line is positive. The "triple point" is indicated by T_p on the phase diagram. This is a unique pressure and temperature condition in which all three phases (solid, liquid, and gas) coexist. For water, it is at 0.01°C and 0.006 atm.

We focus on the liquid/gas equilibrium line. The symbol C_p denotes the critical point at which the boundary line between the liquid and gas phases terminates. At C_p these two phases become indistinguishable and form a single phase, which is known as a supercritical fluid. The C_p for water occurs at 374°C and 218 atm. The region in the phase diagram between the dashed lines, which start at C_p, is for the supercritical fluid phase. It occurs along or above and to the right of the dashed lines, which represent the values for the critical temperature and pressure.

Now we contemplate the path toward C_p, by following the liquid/gas equilibrium line, and then to the supercritical fluid region, after the equilibrium line stops. We notice that the high temperatures that are achieved are in conjunction with an increase in pressure. The approach to the critical point has some

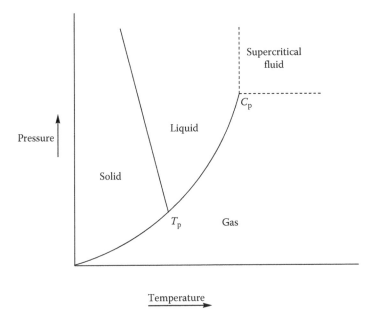

FIGURE 5.2 An approximate and simplified phase diagram for water, shown on a nonuniform scale (free scale).

TABLE 5.1
Terms and Definitions Associated with Phase Diagrams and Superheated Water

Terms	Definitions
Phase diagram	A diagram that shows the regions of pressure and temperature at which various phases of a substance are thermodynamically stable
Phase boundaries	The lines that separate the regions in the phase diagram; they show the values of pressure and temperature at which two phases coexist in equilibrium
Boiling point of water	100.0°C at 1 atm
Critical temperature, T_c	A temperature at which the density of vapor (gas) of a substance is equal to that of its liquid, and the surface between the two phases disappears
Critical pressure, p_c	A pressure at the critical temperature
Critical point	A point on the phase diagram at which the liquid and vapor (gas) phases merge together into a single phase
Critical point for water	374°C and 218 atm
Supercritical fluid	A single uniform phase at and above the critical point
Triple point, T_p	A point on the phase diagram at which the solid, liquid, and vapor (gas) boundaries coexist in equilibrium

(Continued)

TABLE 5.1 (*Continued*)
Terms and Definitions Associated with Phase Diagrams and Superheated Water

Terms	Definitions
Triple point for water	0.01°C and 0.006 atm
HTW (high temperature water)	Water at temperature above its normal boiling point (100°C at 1 atm) (Savage and Robacz, 2010).
SCW (supercritical water)	Water above its critical temperature (374°C) and pressure (218 atm). Distinct gaseous phase and liquid phases no longer exist; there is only a single phase
NCW (near critical water)	Water at temperatures of 200°C–300°C (Liotta et al., 2007).
UCST (upper critical solution temperature)	The temperature above which the organic substrate and water are miscible in all proportions (Liotta et al., 2007)
Gas vs. vapor, as related to the phase diagrams	These two terms are often used interchangeably. For example, Atkins and de Paulo (2006), among others, use vapor. Some other sources use gas (e.g., Hill et al., 2010). Students may have seen either gas or vapor terms on the phase diagrams in their beginning chemistry textbooks, and should not be confused to see that a different term may be used in other textbooks.

Source: Atkins, P. and de Paulo, J., *Atkins' Physical Chemistry*, 8th edn., W. H. Freeman and Company, New York, 2006, pp. 117–121, unless otherwise noted. With permission.

well-defined temperature and pressure regions that are associated with the various types of superheated water, such as near critical, critical, and supercritical. The terms associated with superheated water are more closely defined in Table 5.1. Section 5.2 will explore these regions in more detail and will show that the familiar properties of ambient water change dramatically when water becomes superheated.

5.2   SUPERHEATED WATER AS A GREEN REACTION MEDIUM

In this section, we first discuss the special properties of superheated water. Then we point out the innovative aspects of using superheated water as a green medium for performing organic reactions.

5.2.1   PROPERTIES OF SUPERHEATED WATER

Some properties of water, such as density, dielectric constant, and K_w, are different for ambient water (at 25°C and 1 atm) compared to high-temperature water (HTW) and supercritical water (SCW). This is shown in Table 5.2, which represents selected data from Savage and Rebacz (2010).

TABLE 5.2

Selected Properties of Water under Different Temperatures and Pressures

Property	Ambient Water	HTW	SCW	
Temperature (°C)	25	250	400	400
Pressure (bar)	1	50	250	500
Density (kg/m³)	997	800	170	580
Dielectric constant	78.5	27.1	5.9	10.5
K_w	10^{-14}	$10^{-11.2}$	$10^{-19.4}$	$10^{-11.9}$

Source: Savage, P.D. and Rebacz, N.A.: Water under extreme conditions for green chemistry. *Handbook of Green Chemistry.* 2010. Vol. 5. 331–361. Li, C-J., ed. Copyright Wiley-VCH GmbH & Co. KGaA.

Note: 1 atm = 1.01325 bar

In the discussion of the entries of the table, we make an approximation that 1 bar = 1 atm. Inspection of the values from this table shows the following: Compared to the ambient liquid water at 25°C and 1 atm, HTW, at 250°C and 50 atm, has decreased values of density and dielectric constant, but a substantially increased value of K_w. For SCW at 400°C, these properties vary with pressure, and thus can be modulated by the changes in pressure. For example, SCW at 250 atm shows a dramatic decrease in K_w compared to the ambient water, whereas at 500 amt K_w is substantially increased.

The increased K_w of HTW and SCW under specific conditions means that water becomes more acidic (and basic) than the ambient water. If a reaction requires acid catalysis at ambient conditions, in many cases it will occur in superheated water with no external catalysts, because the H^+ provided by the superheated water is sufficient. This would be true also for the base-catalyzed reactions. This has been amply demonstrated by numerous experiments in superheated water.

The sensitivity of the water parameters to the variations in temperature and pressure has also been reported by Liotta et al. (2007), who use the term near-critical water (NCW) to cover water from 200°C to 300°C. These authors point out that NCW is also referred to as HTW, "hot water," "subcritical water," and "near-subcritical water," in different literature sources. We retain the NCW term in the examples below which are from Liotta et al. (2007).

Under the near-critical conditions, water loses ~55%–60% of its hydrogen-bonding network as the temperature is increased from 25°C to 300°C. The decrease in hydrogen bonding causes a decrease in the dielectric constant. Thus, the dielectric constant at 300°C is ~20, or a 75% reduction from the value under ambient conditions. The NCW value for the dielectric constant corresponds most closely to a moderately polar solvent, such as acetone (dielectric constant of 21.4 at 25°C). This enables enhanced solubility of nonpolar organic species in NCW.

In NCW, the K_w increases by as much as 3 orders of magnitude, from 10^{-14} at 25°C to nearly 10^{-11} at ~250°C, where a maximum occurs. This allows a fine-tuning of the K_w by the pressure and temperature changes, to meet specific demands for the acid/base catalysis of various organic reactions.

5.2.2 INNOVATIVE ASPECTS OF USING SUPERHEATED WATER AS A GREEN REACTION MEDIUM

The basic knowledge of superheated water has been applied only relatively recently to green chemistry. This happened when the innovative breakthrough was achieved, in which the examination of the possibilities of superheated water as a medium for green chemistry reactions found literally a green gold mine. These possibilities are more than compensated for other aspects of the superheated water medium which are not green. The latter include energy requirements to achieve high temperatures and pressures, high cost of the special reaction containers and other equipment, and inherent hazard of pressurized high-temperature systems.

The green innovations as related to the use of superheated water are described in the recommended references. A particularly clear description by Liotta et al. (2007) is given in the text that follows. It also reiterates the important properties of the superheated water, which we have already covered in this chapter.

Superheated water behaves as acetone in terms of dissolving organic compounds. Acetone is as close to the universal solvent as we know it. Thus, all of the sudden, water, which is typically quite poor in dissolving organic compounds, becomes a terrific solvent. As we know, water is inherently a benign solvent. However, there is more. We have shown that the superheated water has both acid and base catalytic properties. A maximum of K_w occurs in NCW at ~250°C. This provides us with a green possibility of performing organic reactions that require acid or base catalysis without adding any mineral acids or bases. The enormous increase in concentrations of H^+ and OH^- ions provides a green catalytic medium. After the reaction is done, simple cooling of the mixture to the ambient conditions restores the neutral conditions, without addition of any external acids or bases that would be normally needed for neutralization. This also enables the ease of separation, because there are no catalysts to be removed or recovered. The workup procedure is simplified because no neutralization is required. Waste reduction is achieved because there is no salt to be disposed (Liotta et al., 2007).

The innovative factor for the use of superheated water in green chemistry is also the use of an inherently nongreen system for green applications, in a way to achieve an overall positive green effect. This requires complex thinking. If we think linearly and focus only on the undesirable properties of superheated water, we would not even consider superheated water as useful for green chemistry.

5.3 ORGANIC REACTIONS IN SUPERHEATED WATER: AN OVERVIEW

Numerous publications such as reviews, chapters, and articles, among others, have described organic reactions in superheated water. These reactions run the gamut from organic synthesis, degradation, and hydrolysis to biomass liquefaction and total oxidation of pollutants. Because these reactions are relatively new, much work had to be done on the basic aspects of their reaction mechanisms, synthetic feasibility, and general scope. The green chemistry applications of these reactions have to be balanced against energy expenditures that are high for the subcritical water and SCW conditions. Considerations must also be given for the potentially aggressive nature of superheated water, particularly if halide ions and oxygen are present. This leads to the strong corrosion of stainless steel containers, among other problems (Kruse and Dinjus, 2012b).

Many studies use simple model compounds to investigate organic reaction mechanisms. Often, a comparison can be made of reaction outcomes between well-known reactions that occur under standard conditions and those performed in superheated water. Green analysis may reveal advantages, but also some disadvantages compared to traditional reaction media. Possibilities also exist for new reaction pathways, because the reaction mechanism sometimes is changed in superheated water.

Table 5.3 is an overview of some illustrative reactions. The material is selected to include the reactions that are familiar to beginning organic chemistry students. More advanced students and other readers will find a comprehensive coverage of the

TABLE 5.3
Overview of Organic Reactions in Superheated Water

Reaction Type/Name	Examples of Substrates and Resulting Products	References
Hydrolysis	Amides (to corresponding carboxylic acid), nitriles (to corresponding amides and carboxylic acids), esters (to corresponding acids and alcohols), aryl ethers (to phenols)	Kruse and Dinjus (2012b)
Rearrangements: Beckmann, Pinacol, and Claisen	Beckmann: oximes (rearranged to lactams); Pinacol (pinacol rearranged to pinacolone); Claisen: allyl phenyl ether (rearranged to 2-allyl phenol)	Savage and Rebacz (2010) Liotta et al. (2007) Kruse and Dinjus (2012b)
Denitration (removal of nitro group)	4-Nitroaniline (to aniline) and N,N-dimethyl-4-nitroaniline (to N,N-dimethylaniline)	Liotta et al. (2007)
Alkylations (Friedel–Crafts type)	Phenol and t-butyl alcohol (to give phenol with t-butyl group in the ring)	Liotta et al. (2007) Avola et al. (2013)
Condensations (Aldol type, such as Claisen–Schmidt and related reactions)	Benzaldehyde and acetone, benzaldehyde and 2-butanone (to give dehydrated aldol)	Liotta et al. (2007) Savage and Rebacz (2010) Avola et al. (2013)
Additions (Diels–Alder type)	Cyclopentadiene and dimethyl maleate to give the Diels–Alder adduct	Avola et al. (2013)
Dehydrations	Alcohols to alkenes, various examples	Avola et al. (2013)
Dehalogenation of alkyl and aryl halides	Dechlorination of 1-chloro-3-phenylpropane, 2-chlorotoluene, 4-chlorophenol, among other examples	Savage (1999)
Oxidations	Toluene to benzaldehyde	Avola et al. (2013)

Note: No externals catalysts, acids, or bases are necessary. By-products may be obtained in some cases.

reactions in superheated water in the recommended references. Chemical equations for selected reactions from Table 5.3 are shown in Figure 5.3.

Some of the yields of the reactions are poor, as can be seen from Figure 5.3. This is mostly due to the substantial side products that form.

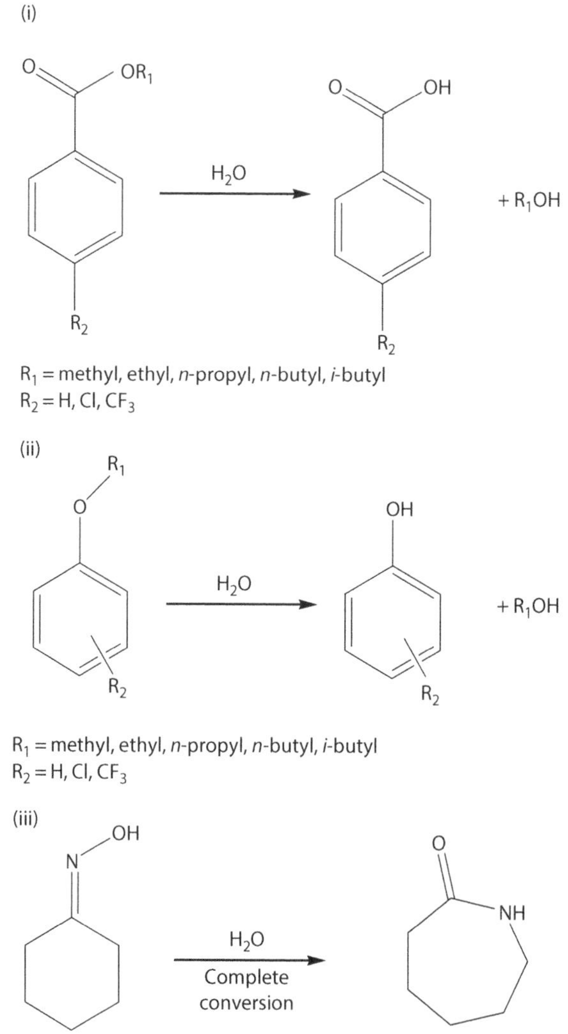

(i)

R_1 = methyl, ethyl, n-propyl, n-butyl, i-butyl
R_2 = H, Cl, CF_3

(ii)

R_1 = methyl, ethyl, n-propyl, n-butyl, i-butyl
R_2 = H, Cl, CF_3

(iii)

FIGURE 5.3 Chemical equations for selected reactions that occur in superheated water from Table 5.3. i, hydrolysis of an ester (From Liotta, C. L. et al., Reactions in near-critical water, in *Organic Reactions in Water: Principles, Strategies and Applications*, Lindström, U. M., ed., Blackwell Publishing, Oxford, pp. 256–300, 2007.); ii, hydrolysis of an ether (From Liotta, C. L. et al., Reactions in near-critical water, in *Organic Reactions in Water: Principles, Strategies and Applications*, Lindström, U. M., ed., Blackwell Publishing, Oxford, pp. 256–300, 2007.); iii, Beckmann rearrangement. (From Kruse, A. and Dinjus, E., Sub- and supercritical water, in *Water in Organic Synthesis*, Kobayashi, S., ed., Thieme, Stuttgart, Germany, pp. 749–771, 2012b.) *(Continued)*

(iv)

(v)

(vi)

FIGURE 5.3 (Continued) Chemical equations for selected reactions that occur in superheated water from Table 5.3. iv, Claisen rearrangement (From Kruse, A. and Dinjus, E., Sub- and supercritical water, in *Water in Organic Synthesis*, Kobayashi, S., ed., Thieme, Stuttgart, Germany, pp. 749–771, 2012b. With permission.); v, Friedel–Crafts alkylation (From Avola, S. et al., Organic chemistry under hydrothermal conditions, *Pure. Appl. Chem.*, 85, 89–103, 2013; Liotta, C. L. et al., Reactions in near-critical water, in *Organic Reactions in Water: Principles, Strategies and Applications*, Lindström, U. M., ed., Blackwell Publishing, Oxford, pp. 256–300, 2007. With permission.); vi, Claisen–Schmidt condensation. (From Avola, S. et al., Organic chemistry under hydrothermal conditions, *Pure. Appl. Chem.*, 85, 89–103, 2013; Liotta, C. L. et al., Reactions in near-critical water, in *Organic Reactions in Water: Principles, Strategies and Applications*, Lindström, U. M., ed., Blackwell Publishing, Oxford, pp. 256–300, 2007. With permission.)

This chapter is concluded by noting that there are many practical green applications of reactions in superheated water. Notably, hydrolytic decomposition of various polymers can be achieved to give more environmentally friendly products, or for the purpose of recycling. For example, polyethylene terephthalate, a polymer that is used in plastic soft drink bottles, can be hydrolyzed "quantitatively" back to the starting materials in less than 1 h in superheated water. Other polymers, such as polyesters, polyamides (such as nylon), polycarbonates, polyethers, polyurethanes, and phenolic resins, are likewise hydrolyzed. For example, polyurethane foams can be hydrolyzed to reusable diamine and glycols. Polyacrylonitrile can be hydrolyzed to low-molecular-weight water-soluble products. This hydrolysis gives ammonia instead of the toxic hydrogen cyanide, which is formed by thermolysis (Siskin and Katritzky, 2001; da Silva, 2007).

Hydrolysis in superheated water is also used to remove sulfur or nitrogen from organic compounds typically found in coal or oil, thus reducing their environmental pollution (Kruse and Dinjus, 2012b). We shall revisit some of these reactions in the forthcoming chapters.

REVIEW QUESTIONS

5.1 Consider the phase diagram of water shown in Figure 5.2. Based on this diagram, explain how the kitchen pressure cookers work, and why do we use them.

5.2 Evaluate the following features of superheated water as a reaction medium in terms of the most relevant principles of green chemistry, and label them as "green" or "not green":
   a. Superheating water
   b. High temperature and pressure
   c. Water as a solvent
   d. External acid or base catalysts are not needed
   e. Neutralization of the acidic or basic conditions occurs without addition of external base or acid, respectively
   f. There will be no salt waste.

5.3 Friedel–Crafts alkylations give quite poor yields in the superheated water, as shown in the example from Figure 5.3. Why do scientists then bother exploring these reactions in superheated water?

ANSWERS TO REVIEW QUESTIONS

5.1 The increase in pressure in the pressure cooker results in the higher cooking temperatures. Thus, we can cook faster.

5.2 a. Principle 6 (requirement for energy minimization) is not fulfilled. Principle 12 (reference to the form of substance that could cause explosions) is not fulfilled. Thus, (a) is not green.
   b. Same as (a); thus, (b) is not green.
   c. Principle 5 (innocuous solvent) is fulfilled; thus, (c) is green.
   d. Principle 9 (about catalysts) is fulfilled; thus, (d) is green.
   e. Principle 5 (eliminate the use of auxiliary substances) is fulfilled; thus, (e) is green.
   f. Principle 1 (prevent waste) is fulfilled; thus, (f) is green.

5.3 Under normal conditions, these reactions usually require Lewis acid catalysts, such as $AlCl_3$, or protic acids, such as sulfuric or phosphoric acids. These acids must be neutralized and separated from the product in the workup and isolation procedure, and the salts from neutralization need to be disposed. Further, $AlCl_3$ produces $Al(OH)_3$, which goes into the landfills. In the superheated water, specifically NCW, water acts as an acid catalyst, so no external catalyst is needed. A simple cooling to the ambient temperature brings the pH to neutral, and thus, no neutralization is required. In sum, the reaction in superheated water provides its own green catalysis. This is a major benefit that provides an impetus to search for solution for unsatisfactory yields.

REFERENCES

Akiya, N. and Savage, P. E. (2002). Roles of water for chemical reactions in high-temperature water, *Chem. Rev.*, 102, 2725–2750.

Atkins, P. and de Paulo, J. (2006). *Atkins' Physical Chemistry*, 8th edn., W. H. Freeman and Company, New York, pp. 117–121.

Avola, S., Guillot, M., da Silva-Perez, D., Pellet-Rostaing, S., Kunz, W., and Goettmann, F. (2013). Organic chemistry under hydrothermal conditions, *Pure. Appl. Chem.*, 85, 89–103.

Boero, M., Ikeshoji, T., Liew, C. C., Terakura, K., and Parrinello, M. (2004). Hydrogen bond driven chemical reactions: Beckmann rearrangement of cyclohexanone oxime into ε-caprolactam in supercritical water, *J. Am. Chem. Soc.*, 126, 6280–6286.

Comisar, C. M., Hunter, S. E., Walton, A., and Savage, P. E. (2008). Effect of pH on ether, ester, and carbonate hydrolysis in high-temperature water, *Ind. Eng. Chem. Res.*, 47, 577–584.

da Silva, F. de C. (2007). Organic reactions in superheated water, *Res. J. Chem. Environ.*, 11, 72–73.

Hill, J. W., McCreary, T. W., and Kolb, D. K. (2010). *Chemistry for Changing Times*, 12th edn., Prentice Hall, New York, p. 151 (about supercritical fluids and the green applications of supercritical CO_2).

Ikushima, Y., Hatakeda, K., Sato, O., Yokoyama, T., and Arai, M. (2000). Acceleration of synthetic organic reactions using supercritical water: Noncatalytic Beckmann and Pinacol rearrangements, *J. Am. Chem. Soc.*, 122, 1908–1918.

Katritzky, A. R., Nichols, D. A., Siskin, M., Murugan, R., and Balasubramanian, H. (2001). Reactions in high-temperature aqueous media, *Chem. Rev.*, 101, 837–892.

Kruse, A. and Dinjus, E. (2012a). Sub- and supercritical water, in *Water in Organic Synthesis*, Kobayashi, S., ed., Thieme, Stuttgart, Germany, p. 750.

Kruse, A. and Dinjus, E. (2012b). Sub- and supercritical water, in *Water in Organic Synthesis*, Kobayashi, S., ed., Thieme, Stuttgart, Germany, pp. 749–771.

Lancaster, M. (2002). *Green Chemistry: An Introductory Text*, Royal Society of Chemistry, Cambridge, pp. 135–154.

Li, C.-J. and Chan, T.-H. (2007). *Comprehensive Organic Reactions in Aqueous Media*, 2nd edn., Wiley, Hoboken, NJ, pp. 9–13.

Liotta, C. L., Hallett, J. P., Pollet, P., and Eckert, C. A. (2007). Reactions in near-critical water, in *Organic Reactions in Water: Principles, Strategies and Applications*, Lindström, U. M., ed., Blackwell Publishing, Oxford, pp. 256–300.

Savage, P. D. and Rebacz, N. A. (2010). Water under extreme conditions for green chemistry, in *Handbook of Green Chemistry*, Vol. 5, Li, C.-J., ed., Wiley-VCH, Weinheim, Germany, pp. 331–361.

Savage, P. E. (1999). Organic reactions in supercritical water, *Chem. Rev.*, 99, 603–621.

Siskin, M. and Katritzky, A. R. (2001). Reactivity of organic compounds in superheated water: General background, *Chem. Rev.*, 101, 825–835.

6 Green "Solventless" Organic Reactions

The remarkable versatility and success of using solventless reactions to prepare several classes of compounds demonstrates that this methodology has an important place in the toolbox arsenal for Green Chemistry.

Gareth W. V. Cave, Colin L. Raston, and Janet L. Scott (2001a)

…it is remarkable that chemists still carry out their reactions in solution, even when a special reason for the use of solvent cannot be found.

Koichi Tanaka and Fumio Toda (2000)

LEARNING OBJECTIVES

The learning objectives for this chapter are as follows.

Learning Objectives	Section Numbers
Learn that organic reactions may not need a solvent to occur	6.1
Learn about solventless reactions	6.2
Learn about solventless reactions that are based on the melting behavior of the reaction mixtures	6.3
Revisit the familiar concepts of melting point and eutectic point, and the phase diagrams to better understand solventless reactions	6.4
Learn about challenges of solventless reactions to green chemistry	6.5
Learn about various examples of solventless reactions	6.6

6.1 ANOTHER GREEN CHEMISTRY PARADIGM SHIFT: ORGANIC REACTIONS MAY NOT NEED A SOLVENT TO OCCUR

Throughout the history of organic chemistry, the use of organic solvents as a reaction medium was considered necessary, unavoidable, and thus mandatory. Chemists believed that solvents increase mobility of the reactants' molecules. Such mobility then facilitates molecular collisions in general and successful collisions in particular, because the latter usually require specific orientations of molecules. Because the organic reactions were always done in solvents, a dogmatic view was established that this practice is correct and should not be questioned. However, with the birth of the green chemistry movement, the negatives about the use of organic solvents as a reaction medium (or in any other capacity, such as for separation and purification of the reaction products) came to light. Many organic solvents are toxic to humans and other life, and may be hazardous to the environment. In addition, they may be flammable or explosive. Notably, many solvents are volatile organic compounds (VOCs), which can easily escape to the environment. Containment of VOCs is often difficult, costly, and risky. With a rapid growth of organic chemistry and its industrial applications, pollution with organic solvents in general, and especially with VOCs, became a rather serious problem. Pollution cleanup became quite costly, and it was not even economically feasible in many cases. We have discussed many of these problems in Chapter 1. As a consequence of these negative features of many organic solvents, the green chemistry founders and practitioners asked a bold question: "Can we run organic reactions without solvents?" As a search for the answer was conducted, the paradigm that organic solvents are necessary for running reactions started crumbling. Ultimately, green chemistry overturned the old paradigm and has introduced a new one: not only can organic reactions be run without solvents, but often such reactions gave unexpected benefits for the reaction outcomes, such as faster reactions, higher yields, and improved selectivity. The green advantages are numerous. In addition to the elimination of solvents, and thus the need for their disposal or recycling, there is a reduction in pollution and a simplification of chemical processes. We shall provide specific examples in the forthcoming chapters on applications of solventless reactions to various industries.

6.2   DEFINITION OF SOLVENTLESS REACTIONS AND RELATED NOMENCLATURE

The literal meaning of "solventless" is "without solvent." As we have learned from Section 6.1, many organic reactions have been shown to occur without organic solvents and often give better results than those occurring in a solvent. There are many different ways solventless reactions can occur. Mechanisms of these reactions differ and many are still under study. Further, different investigators use different terms for the same or similar soventless reaction conditions. The definition and a proper use of these terms are important, especially for the literature searches, which often depend on the use of very specific key words.

The terms that are often used interchangeably with "solventless" are "solvent-free," "in the absence of solvents," and "neat." The latter term is used to cover the reactions between neat reagent combinations, such as gas/solid, solid/liquid, liquid/liquid, and solid/solid (Varma and Ju, 2005).

Our main focus in this chapter is on solventless solid/solid reactions that are affected by mixing or grinding of two or more reactants, which results in the melting of the reaction mixture. Such reactions are among the most popular green chemistry laboratory experiments. In Chapter 7, we shall address organic reactions in the solid state more comprehensively, and will discuss their mechanisms, which often blur the lines between solid/solid and solventless reactions. Solid/liquid and liquid/liquid reactions are easier to understand, because the liquid phase acts as a "solvent."

6.3 SOLVENTLESS REACTIONS BASED ON THE MELTING BEHAVIOR OF REACTION MIXTURES: INNOVATIVE ASPECTS OF THIS METHOD

As an introduction to the mechanism by which solid/solid solventless reactions occur, we first review the melting behavior of organic solids, which is at the core of this mechanism.

One of the basic organic laboratory skills is a determination of melting point of solid organic compounds, for the purpose of their identification and evaluation of their purity. Students learn that if the melting point of a compound is lower than its literature value, and the melting occurs over a wide temperature range, the compound is impure. In contrast, pure compounds give sharp melting points, which do not deviate from the literature values by more than 1°C–2°C. For impure compounds, a large melting range could be partially explained by residual solvents that were used for isolation or recrystallization, unreacted starting materials, and side products, among others.

The melting point can also be used for characterization of unidentified organic compounds ("the unknown"), via determination of a so-called mixed melting point. In this procedure, one prepares approximately a 1:1 mixture of the unknown compound with a known compound that one believes the unknown could be. Then one takes a melting point of such a mixture. If the melting point of the mixture is the same as for the unknown, then the unknown compound is identical to the suspected known compound. If the melting point of the mixture is substantially lowered and broadened, then the guess about the identity of the unknown is incorrect.

The lowering of the melting point by impurities appears to be such a common, well-known, and often observed phenomenon that one wonders if there is anything that one can add to the usefulness of this basic and routine organic laboratory tool. However, green chemistry researchers have found a major innovative application of this principle, which resulted in the discovery of solventless reactions (e.g., Raston and Scott, 2000; Cave et al., 2001b; Palleros, 2004; Raston, 2004; Touchette, 2006). The description of this innovation as it follows. Suppose that we do not have a desired solid compound A mixed with the annoying solid impurity B, but, instead, that solids A and B are the reactants in a reaction that will give a desired product. If A and B act as each other's impurities to the extent that the melting point of the mixture is lowered below room temperature, the mixture then becomes a liquid. The liquid phase then will act as a "solvent." The reaction that occurs in such a melted state is "solventless." This is very clever and ingenious. As in the case of many other innovations, we are prone to ask ourselves: "This was rather simple and obvious, why did I not come up with this?"

Figure 6.1 shows a classical example of a solventless reaction (Touchette, 2006). The reaction shown in the figure is pedagogically friendly, and it dramatically illustrates the principle of the solventless reaction. Both starting materials, *o*-vanillin and *p*-toluidine, are low melting solids; they melt at 40°C–42°C and 44°C–45°C, respectively. Both are nearly colorless. When the two solids are put in contact, even before they are thoroughly mixed and ground, a bright orange color is observed, which signifies the formation of the imine product. This makes a very nice demo. Upon mixing, the reaction progresses quickly. It is done in about 5 min. The reaction is quite

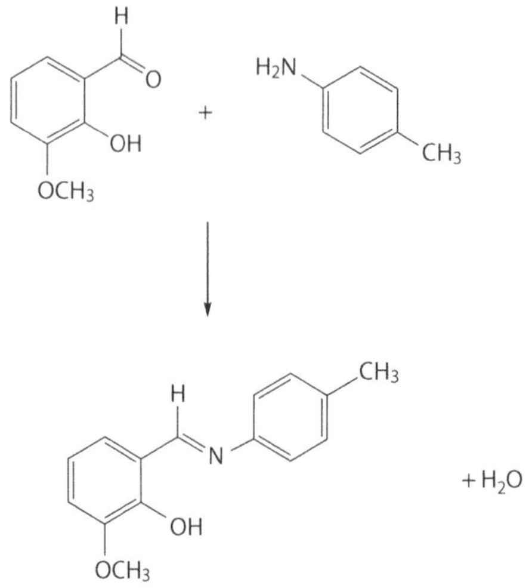

FIGURE 6.1 Soventless reaction between *o*-vanillin and *p*-toluidine to give the corresponding imine. (From Touchette, K. M., Reductive amination: A remarkable experiment for the organic laboratory, *J. Chem. Ed.*, 83, 929–930, 2006. With permission.)

exothermic. Water that is formed as a side product is rapidly evaporated by the heat that is generated in the reaction. After 5 min, one is left only with the solid orange imine.

In another scenario, the reaction components upon mixing melt into a liquid as before. However, at this point, a reaction catalyst is added and the reaction now starts to occur and yield product(s). This is illustrated in another popular solventless reaction, an aldol condensation, which is shown in Figure 6.2.

The procedure for this reaction calls for mixing and grinding together the two reactants, which will then melt to give an oil. Next, solid NaOH is added to catalyze

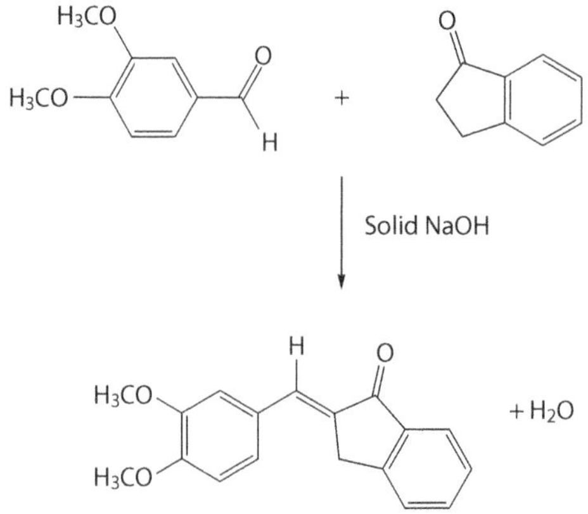

FIGURE 6.2 Solventless aldol condensation between 3,4-dimethoxybenzaldehyde and 1-indanone, with solid NaOH as a catalyst. (From Doxsee, K. M. and Hutchinson, J. E., *Green Organic Chemistry, Strategies, Tools, and Laboratory Experiments*, Thomson Brooks/Cole, Toronto, Ontario, Canada, pp. 115–119, 2004. With permission.)

the reaction. The reactants are again low melting solids. The aldehyde melts at 41°C–42°C and the ketone at 45°C–48°C.

Notice that in both the examples given above, the molar ratio of reactants is 1:1, and the melting points of the reactants are low. A question arises about applicability of such solventless method to reactions in which a different stoichiometric ratio of reactants is required and in which the melting points of the reactants may be much higher. To evaluate the feasibility of such reactions as solventless, we need to learn more about the relationship between the melting point and the composition of the mixture. This is addressed in Section 6.4.

These and other solventless reactions are green experiments that are often performed in the undergraduate organic laboratories (see Chapter 13).

6.4   A NEW LOOK AT THE FAMILIAR CONCEPTS OF MELTING POINT AND EUTECTIC POINT, AND THEIR RELEVANT PHASE DIAGRAMS

The phenomenon of the decrease of a melting point by an impurity that is present is best understood by an examination of a melting point–composition diagram for a mixture of two components, A and B (thus a binary mixture). Figure 6.3 represents such a diagram. It should be noted that a mixture comprising more than two compounds may also be considered, but the corresponding diagram would be more complicated. Although A and B in principle may react upon melting, as shown in Figure 6.1, we shall exclude this possibility for now. This will simplify the diagram;

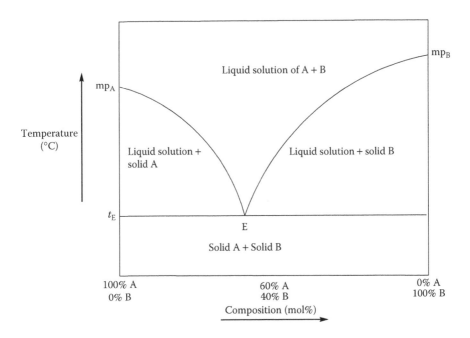

FIGURE 6.3   Melting point–composition diagram for a binary mixture composed of A and B.

otherwise, additional component(s) corresponding to the product(s) would have to be included. What we are considering here would be applicable to the case shown in Figure 6.2 before the addition of the catalyst.

The diagram in Figure 6.3 shows a case in which a mixture of A and B, whose structures and melting points are not specified, melts sharply at the point *E*, at a specific molar composition of A and B, which is here assigned arbitrarily as 60%:40%. The point *E* is called a eutectic point. This term comes from the Greek word for "easily melted." A solid with the eutectic composition melts at the lowest temperature of any mixture. The melting occurs without change of composition. Below the eutectic temperature, there is a solid phase.

The eutectic point can occur at any ratio between two compounds, which depends on their structures and melting points. It should be noted that some mixtures of compounds may show no eutectic point or a more complex behavior, in which there is more than one eutectic point.

The diagram shown in Figure 6.3 is valid only under a restriction that A and B are soluble in the liquid phase. A more in-depth coverage of this phenomenon is given by Pedersen and Myers (2011).

The melting point is recorded as a range between the first appearance of a liquid and the complete melting. However, the application of the diagram from Figure 6.3 would not be realistic in this respect. One needs to use Figure 6.4, which includes the experimental curve at which the melting starts (Harding, 1999).

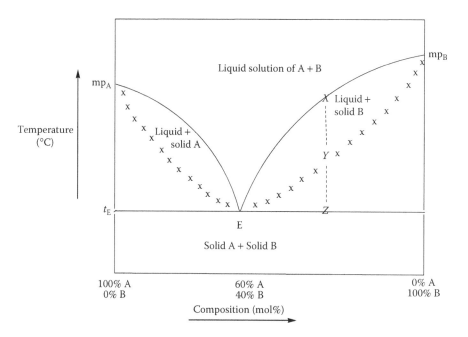

FIGURE 6.4 The experimental curve (xxx) at which the melting starts. The line Z–X is the theoretical melting range, and the line Y–X is the observed range. (From Harding, K. E., Melting point range and phase diagrams—Confusing laboratory text-book descriptions, *J. Chem. Ed.*, 76, 224–225, 1999. With permission.)

From Figure 6.3, the melting range would start at the t_E horizontal line and would end at the curve that denotes a boundary with the pure liquid phase. This would give an unrealistically broad and incorrect melting range. This is illustrated in Figure 6.4 as the line joining points Z and X. Further, and unrealistically, the melting range would become increasingly broad as the purity of the compound increases. This is contrary to what is observed. By the introduction of the experimental curve in the diagram in Figure 6.4, this problem is eliminated. Thus, the melting is observed to start not at Z, but at Y, and it finishes at X. The observed melting point range is thus Y–Z. It can also be seen that the melting range for an almost pure A or an almost pure B is in a very narrow range and is thus sharp. It should be noted that the diagram from Figure 6.3 is not wrong. It is just that the amount of liquid that is formed at the beginning of melting may be so small that we cannot observe it visually. We observe the melting at the experimental curve.

A practical illustration of the above discussion is shown in some cases of solventless reactions, especially in those that give colored compounds. Although we may not observe any melting at the interface of the two solids, the reaction may nonetheless start there as evidenced by an observed color development (Touchette, 2006; Dolotko et al., 2010). This phenomenon can be easily explained by the experimental difficulty of observing a very small amount of liquid at the beginning of melting, whereas color development can be detected much more readily.

In sum, the melting behavior shown in Figure 6.4 blurs the division between solventless and solid/solid reactions (Rothenberg et al., 2001; Dolotko et al., 2010). The solid/solid reactions may actually occur in a liquid film that is formed by the local melting. When we mix and grind two solids, we create many liquid interfaces, and we may be able to observe the local melting better, but not always.

6.5 GREEN CHEMISTRY CHALLENGES WITH RESPECT TO SOLVENTLESS REACTIONS

Organic solvents, despite showing serious negative characteristics for green chemistry (which we have summarized earlier in this chapter), have numerous, otherwise positive aspects. For example, solvents allow for an easy transfer of starting materials to the reaction vessel, for a gradual (e.g., drop wise) addition of reactants (e.g., for control of competing reactions), for taking aliquots out of a reaction mixture (e.g., for the purpose of following the reaction progress), for absorption and dissipation of heat in cases of exothermic reactions, for diluting the reaction mixture to slow down the reaction if necessary, and for influencing reaction outcome (e.g., promote S_N2 vs. S_N1 reactions).

Next, we address more fully a rather serious problem of exothermic solventless reactions. We have learned from the discussion of the reaction shown in Figure 6.1 (formation of an imine by a solventless condensation reaction) that the reaction is quite exothermic. At a small laboratory-scale experiment (less than a gram of each reactant), water, which is a side product, is rapidly boiled off and lost to evaporation. A scale-up of such a reaction for industrial purposes would be quite challenging, because an uncontrolled evolution of heat without the benefit of a solvent to absorb and dissipate it could be quite hazardous.

The lost positives of the solvents need to be compensated for when the solventless reactions are put into practical use. This represents a serious challenge to green chemistry, but it can be solved by ingenuity. Green solutions need to be tailored to each specific reaction or process.

6.6   EXAMPLES OF SOLVENTLESS REACTIONS

This section gives additional examples of common organic reactions that can be performed as solventless. The examples that we have chosen are those in which the presence of a liquid melt was critical for the reaction to occur as was demonstrated experimentally by Rothenberg et al. (2001). They are presented in Table 6.1.

Figure 6.5 shows the chemical equations for some of the reactions listed in Table 6.1.

There are many solid–solid reactions that are potentially solventless in the sense we describe, namely, requiring a liquid phase to occur. Note that in many cases, the mechanisms of such reactions have not been determined. We shall show a large number of such solid–solid reactions in Chapter 7. They show a great diversity of the types of reactions and an impressive range of synthetic applications.

Various other solvent-free reactions, such as gas/solid, liquid/solid, and liquid/liquid, are also important. Many have been applied successfully to the pharmaceutical and other industries. Specific examples are provided in the forthcoming chapters.

REVIEW QUESTIONS

6.1 In the reaction shown in Figure 6.1, water is a by-product. How would you prove this experimentally?

6.2 In a hypothetical example of a solventless aldol condensation, the two starting materials did not melt upon mixing and grinding. Propose two different ways how you could still perform such a reaction in a solventless manner.

TABLE 6.1
Selected Examples of Solventless Reactions

Reaction Type	Reaction Name
C–C bond formation	Aldol condensation
C–C bond formation	Oxidative coupling of naphthols
Formation of lactones (by reaction of ketones and *m*-chloroperbenzoic acid)	Baeyer–Villiger oxidation
C–N bond formation	Condensation of amines and aldehydes to give azomethines
Ring bromination of aromatics	Reaction of aromatics with *N*-bromosuccinimide
Ether formation	Etherification of alcohols with *p*-toluenesulfonic acids

Source: Rothenberg, G. et al., Understanding solid/solid organic reactions, *J. Am. Chem. Soc.*, 123, 8701–8708, 2001. With permission.

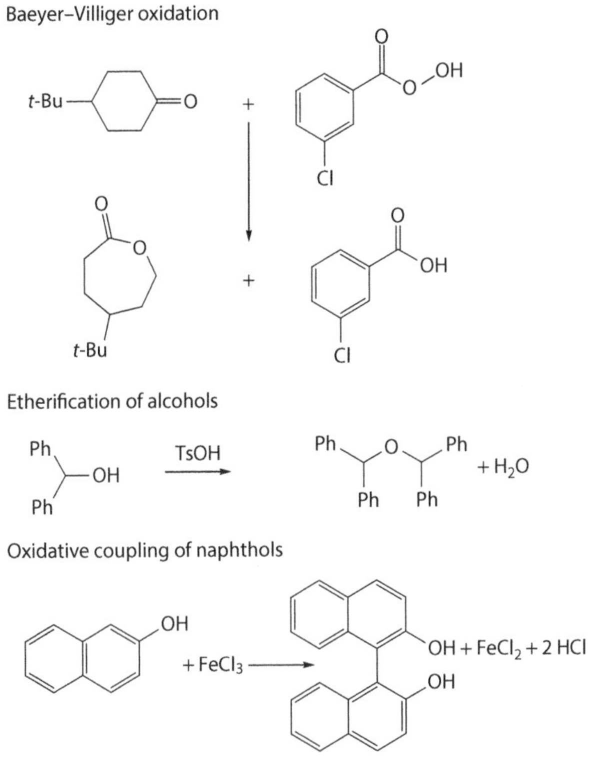

Baeyer–Villiger oxidation

Etherification of alcohols

Oxidative coupling of naphthols

FIGURE 6.5 Chemical equations for selected reactions from Table 6.1.

6.3 Research a solvent-free Cannizzaro reaction, a disproportionation that occurs under the grinding of 2-chlorobenzaldehyde (liquid) with solid KOH to give 2-chlorobenzoic acid (solid) and 2-chlorobenzyl alcohol (solid) (see Dicks, 2009; Phonchaiya et al., 2009; your organic laboratory textbook may also have this reaction). Because 2-chlorobenzaldehyde is a liquid, this reaction is a neat solvent-free reaction. What are the green advantages of the workup and isolation of the products in this reaction?

ANSWERS TO REVIEW QUESTIONS

6.1 An exposure of anhydrous $CuSO_4$, a white solid powder, to the vapors above the orange imine product, or the vapor condensate on the side of the reaction vessel, would result in a formation of the hydrated $CuSO_4$ ($CuSO_4 \cdot 5H_2O$), which is blue.

6.2 The first way would be a gentle heating, which might melt the components. If this does not work, one could add another component, as an intentional impurity, which would not be reactive with the starting materials, the product, or with the base that is needed for catalysis. Such an impurity may help lower the melting point of the mixture below room temperature. This impurity would need to be benign and easily separable and recyclable from the reaction mixture, for the process to be green.

6.3 The workup consists of adding water. No organic solvents are required. The solid 2-chlorobenzyl alcohol is isolated by filtration of the aqueous solution. The solid 2-chlorobenzoic acid forms as a precipitate upon acidification of the mother liquor and is separated by filtration.

REFERENCES

Atkins, P. and de Paula, J. (2006). *Atkins' Physical Chemistry*, 8th edn., W. H. Freeman and Company, New York, pp. 189–194.

Cave, G. W. V., Raston, C. L., and Scott, J. L. (2001a). Recent advances in solventless organic reactions: Towards benign synthesis with remarkable versatility, *Chem. Commun.*, (21), 2167.

Cave, G. W. V., Raston, C. L., and Scott, J. L. (2001b). Recent advances in solventless organic reactions: Towards benign synthesis with remarkable versatility, *Chem. Commun.*, (21), 2159–2169.

Dicks, A. P. (2009). Solvent-free reactivity in the undergraduate organic laboratory, *Green Chem. Lett. Rev.*, 2, 87–100.

Dicks, A. P. (2012). Elimination of solvents in the organic curriculum, in *Green Organic Chemistry in Lecture and Laboratory*, Dicks, A. P., ed., CRC Press, Boca Raton, FL, pp. 69–102.

Dolotko, O., Wiench, J. W., Dennis, K. W., Pecharsky, V. K., and Balem, V. P. (2010). Mechanically induced reactions in organic solids: Liquid eutectics or solid-state processes?, *New J. Chem.*, 34, 25–28.

Doxsee, K. M. and Hutchinson, J. E. (2004). *Green Organic Chemistry, Strategies, Tools, and Laboratory Experiments*, Thomson Brooks/Cole, Toronto, Ontario, Canada, pp. 115–119.

Harding, K. E. (1999). Melting point range and phase diagrams—Confusing laboratory textbook descriptions, *J. Chem. Ed.*, 76, 224–225.

Mohrig, J. R., Alberg, D. G., Hofmeister, G. E., Schatz, P. F., and Hammond, C. N. (2014). *Laboratory Techniques in Organic Chemistry, Supporting Inquiry-Driven Experiments*, 4th edn., W. H. Freeman and Company, New York, pp. 212–213.

Orita, A., Okano, J., Uehara, G., Jiang, L., and Otera, J. (2007). Importance of molecular-level contacts under solventless conditions for chemical reactions and self-assembly, *Bull. Chem. Soc., Jpn.*, 80, 1617–1623.

Palleros, D. R. (2004). Solventless synthesis of chalcones, *J. Chem. Ed.*, 81, 1345–1347.

Pedersen, S. F. and Myers, A. M. (2011). *Understanding the Principles of Organic Chemistry: A Laboratory Course*, Brooks/Cole, Belmont, CA, pp. 60–68.

Phonchaiya, S., Panijpan, B., Rajviroongit, S., Wright, T., and Blanchfield, J. T. (2009). A facile solvent-free Cannizzaro reaction, *J. Chem. Ed.*, 86, 85–86.

Raston, C. L. (2004). Versatility of "alternative" reaction media: Solventless organic syntheses, *Chem. Austr.*, May 2004, pp. 1–13.

Raston, C. L. and Scott, J. L. (2000). Chemoselective, solvent-free aldol condensation reaction, *Green Chem.* April 2000, 2, 49–52.

Rothenberg, G., Downie, A. P., Raston, C. L., and Scott, J. L. (2001). Understanding solid/solid organic reactions, *J. Am. Chem. Soc.*, 123, 8701–8708.

Tanaka, K. and Toda, F. (2000). Solvent-free organic syntheses, *Chem. Rev.*, 100, 1025–1074.

Touchette, K. M. (2006). Reductive amination: A remarkable experiment for the organic laboratory, *J. Chem. Ed.*, 83, 929–930.

Varma, R. S. and Ju, Y. (2005). Solventless Reactions (SLR), in *Green Separation Processes*, Afonso, C. A. M. and Crespo, J. G., eds., Wiley-VCH, Weinheim, Germany, pp. 53–87.

7 Green Organic Reactions in the Solid State

The occurrence of efficient solid-state reactions shows that the molecules reacting are able to move freely in the solid state.

Koichi Tanaka and Fumio Toda (2000a)

LEARNING OBJECTIVES

The learning objectives for this chapter are as follows.

Learning Objectives	Section Numbers
Learn about solid-state reactions	7.1
Become familiar with different types of solid-state reactions	7.2
	7.3
	7.4
Learn about the synthetic scope of solid-state reactions and examine specific examples	7.5

7.1 SOLID-STATE REACTIONS OTHER THAN THOSE THAT OCCUR VIA THE MELTED STATE

In Chapter 6, we have focused on one specific type of solid-state reactions, namely, those that occur via a melted state. Such a state is achieved by the formation of eutectics, which form a liquid phase at the contact point of the solid reagents.

In this chapter, we focus on the type of solid-state reactions in which the solids react directly to form a solid product(s) without intervention of a liquid or vapor phase.

An important question needs to be answered about such reactions: Are molecules in the solid state sufficiently mobile in order to react and give product(s) in noticeable amounts? If such reactions can indeed occur, a further question arises. Would such reactions meet the standards of green chemistry, such as high yields, favorable reaction rates, and desired specificity? The answer to all these questions is affirmative, based on the experimental data.

To answer these questions, three different types of solid-state reactions are discussed. First, we show examples of solid-state photochemical reactions. Some of these reactions give products that are different from those obtained in solution, whereas others give products that cannot be obtained in solution. Thus, photochemical reactions have a unique synthetic potential. Second, examples of microwave-assisted solid-state reactions are illustrated. Typically, these reactions are rapid. Finally, we show examples of molecular rearrangements that can occur in the solid state. This type of reaction dramatically illustrates the mobility of molecules in the solid state.

7.2   SOLID-STATE PHOTOCHEMICAL REACTIONS

It has been known for quite some time that facile photochemical reactions do occur in the solid state. Some of these reactions give different products from those obtained in solution (Thomas, 1979). An example is shown in Figure 7.1. Both reactions, in the solid state and in solution, proceed under ultraviolet irradiation.

The example from the figure shows a synthetic utility of a solidstate reaction, which cannot be achieved in solution.

As a second example, Figure 7.2 illustrates a photochemical reaction that occurs in the solid state but does not occur in solution (Doxsee and Hutchinson, 2004).

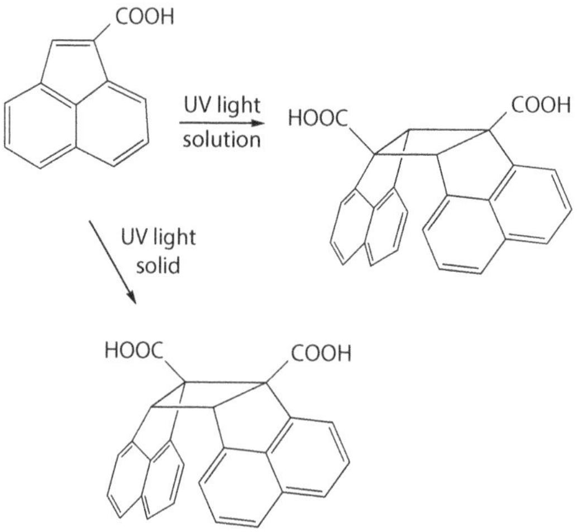

FIGURE 7.1   An example of a facile solid-state photochemical reaction, which gives a product that is different from the product that is obtained in solution. (From Thomas, J. M., Organic reactions in the solid state: Accident and design, *Pure Appl. Chem.*, 51, 1065–1082, 1979. With permission.)

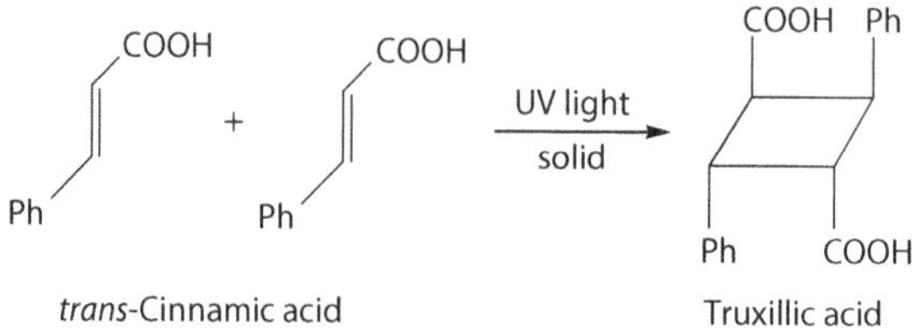

FIGURE 7.2   Solid-state photochemical reaction of *trans*-cinnamic acid, which gives a single stereoisomer of truxillic acid. (From Doxsee, K. M. and Hutchinson, J. E., *Green Organic Chemistry, Strategies, Tools, and Laboratory Experiments*, Thomson Brooks/Cole, Toronto, Canada, pp. 206–210, 2004. With permission.)

In this reaction, only one stereoisomer of truxillic acid is obtained, although five are possible. This product is a result of a "head-to-tail" dimerization.

7.3 MICROWAVE-ASSISTED SOLID-STATE REACTIONS

Microwave heating is considered a green energy method, because it achieves high temperatures quickly, and the reaction times are generally shortened. Many solid-state reactions (and also solid/liquid and liquid/liquid reactions) are facilitated by using microwave heating. Student-friendly laboratory experiments on this subject have been recently published by Leadbeater and McGowan (2013). Also, microwave heating has been used in the student-achievable synthesis of heterocyclic compounds (e.g., Musiol et al., 2006). Heterocyclic compounds have outstanding applications in pharmaceutical industry, because they exhibit a wide range of biological activities, such as antimicrobial, antiparasitic, anticonvulsant, antiviral, and anti-inflammatory activities, among others.

Figure 7.3 shows an example of microwave-assisted synthesis of a heterocyclic compound.

Note the extremely short reaction time, which provides a great green benefit. Also note that the power setting (850 W) is usually specified for these reactions.

7.4 REARRANGEMENT REACTIONS IN THE SOLID STATE

Rearrangement reactions in the solid state are especially of interest, because their occurrence provides support for the mobility of molecules in the solid phase.

A classic example is the rearrangement reaction of pinacol (an alcohol) to pinacolone (a ketone) under acid catalysis conditions. The reaction and its mechanism are shown in Figure 7.4.

This type of rearrangement has been studied extensively, with a variety of alkyl and aryl groups, not just methyl, which was used in the original case shown in Figure 7.4. The reaction is often referred to as a "pinacolone rearrangement," and it covers a broad range of pinacolone analogs.

The mechanism of the pinacolone rearrangement in the solid state was reviewed by Tanaka and Toda (2000b). The example they used is shown in Figure 7.5. The reaction in the solid state is catalyzed by solid TsOH·H$_2$O (a hydrate of p-toluenesulfonic acid, CH$_3$-C$_6$H$_4$-SO$_3$H). This reaction occurs much faster in the solid state than in solution.

Characteristics of solid-state reactions may be explained through models or mechanisms that are formulated from experimental evidence. In the above reaction, atomic force microscopy (AFM) and crystal structure analysis were used to deduce a plausible mechanism for the reaction.

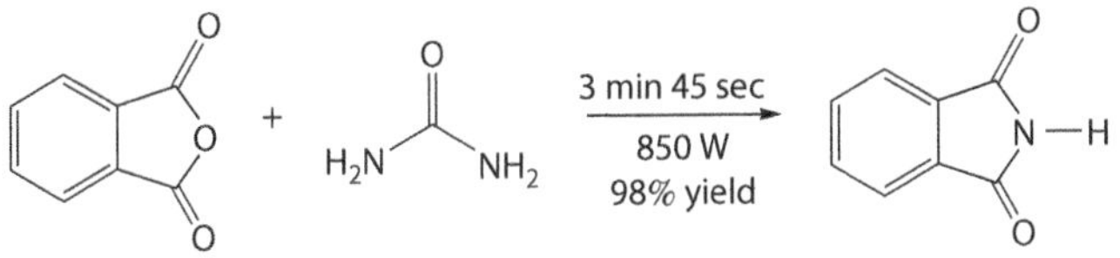

FIGURE 7.3 Microwave-assisted solid-state synthesis of phthalimide, a heterocyclic compound. (From Musiol, R. et al., Microwave-assisted heterocyclic chemistry for the undergraduate organic laboratory, *J. Chem. Ed.*, 83, 632–633, 2006. With permission.)

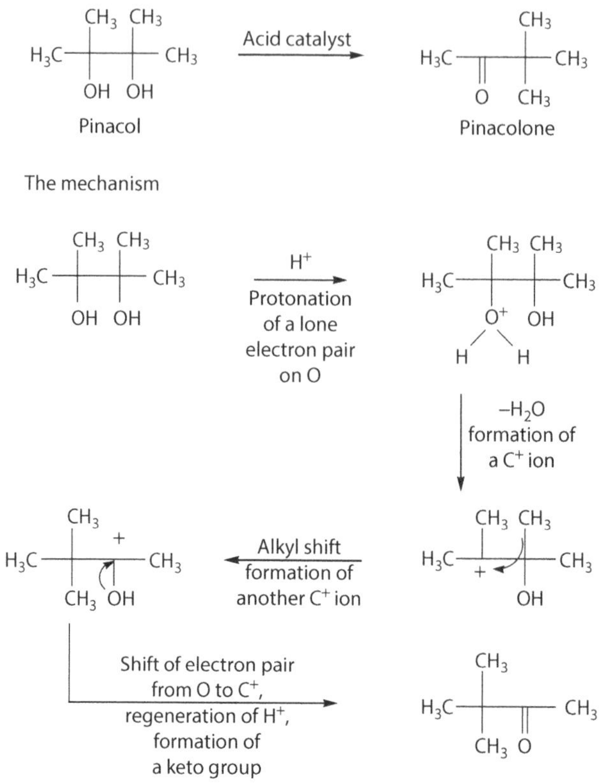

FIGURE 7.4   Pinacol-to-pinacolone rearrangement and its mechanism.

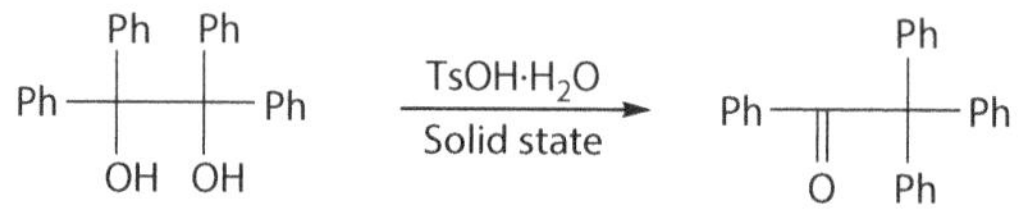

FIGURE 7.5   An example of pinacolone rearrangement in the solid state, which was the subject of a mechanistic study. (From Tanaka, K. and Toda, F., Solvent-free organic synthesis, *Chem. Rev.*, 100, 1025–1074, 2000b. With permission.)

AFM is a very high-resolution type of scanning probe microscopy that provides images of atoms on surfaces. Of interest here were the changes in the surfaces of the pinacol when a tiny crystal of solid-state catalyst, TsOH·H₂O, was placed at a distance of 0.5 mm. The reaction of the solid pinacol was shown to occur only when its OH groups are properly oriented towards the catalyst. The protonation of the OH groups by the catalyst occurs when their oxygen lone electrons are pointing towards the catalyst, while their hydrogens are pointing away. Once the protonation occurs, the reaction starts, and then proceeds by the mechanism shown in Figure 7.4. In the last step, the proton catalyst is regenerated and it reacts with the next molecule. The reaction then proceeds further along the channels in the crystal, leaving behind water that is formed as a by-product (Tanaka and Toda, 2009).

7.5 SYNTHETIC SCOPE OF SOLID-STATE REACTIONS

Solid-state reactions cover a large number of reaction types. Numerous examples of such reactions are reported in the literature (e.g., Tanaka and Toda, 2000b; Tanaka, 2009).

In Table 7.1, we have summarized the types of reactions that have been successfully performed in the solid state and provided selected examples of reactions that students are likely familiar with, or which are reasonably simple. The examples are from the book by Tanaka (2009). This book gives a brief description of the experimental procedure for each reaction listed, as well as the original literature sources. Figure 7.6 shows equations for some examples from Table 7.1.

In conclusion, solid-state reactions are numerous and cover a wide range of reaction types. Many give unique products that are not obtained by reactions in solution. Solid-state reactions can give excellent yields, and can be very fast and selective. We shall explore some practical green applications of these reactions in Chapters 7 and 13.

TABLE 7.1
Selected Types of Reactions That Can Be Performed in Solid State and Some Examples

Type of Reaction	Examples
Reduction	Ketones to alcohols, with $NaBH_4$ (p. 1)
	Imines to amines, with $NaBH_4/H_3BO_3$ (p. 6)
Oxidation	Ketones to lactones, with *m*-chloroperbenzoic acid (Baeyer–Villiger reaction) (p. 11)
	Nitriles to amides, with urea/H_2O_2 (p. 13)
	Alcohols to aldehydes or ketones, with alumina-supported permanganate (p. 15)
Carbon–carbon bond formation	Coupling of acetylenic compounds (propargyl alcohols) to diacetylenic compounds, with $CuCl_2$. 2 pyridine (Glaser coupling) (p. 34)
	Condensation of aromatic aldehydes with cyanoacetamide with base catalyst (Knoevenagel condensation) (p. 81)
	Condensation of aromatic aldehydes with barbituric acids (pp. 82–83)
	Solid-state Diels–Alder reaction (p. 127)
	Solid-state photochemical reactions (pp. 128, 147)
Carbon–heteroatom bond formation	Various reactions in which carbon–nitrogen, carbon–oxygen, and carbon–sulfur bonds are made
Rearrangements	Pinacol to pinacolone (p. 343)
	Benzilic acid rearrangement (of 1,2-diketones to give α-hydroxy carboxylic acids) (pp. 343–344)
	Beckmann rearrangement (of an oxime to an amide) (pp. 346–347)
Other miscellaneous reactions	Elimination, protection/deprotection of functional groups, polymerization

Source: Tanaka, K.: *Solvent-Free Organic Synthesis*. 2009. Copyright Wiley-VCH Verlag GmbH & Co. KGaA.

Note: The numbers in parentheses refer to page numbers from Tanaka (2009).

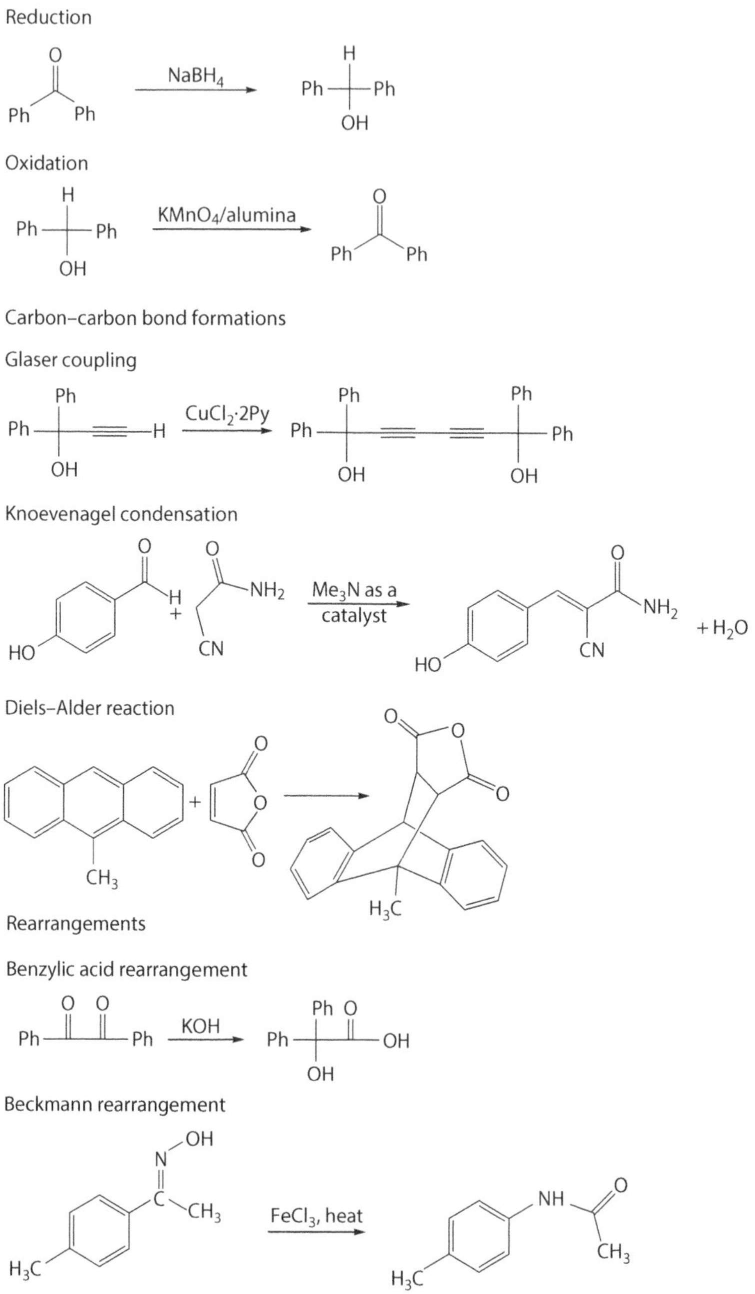

Reduction
NaBH$_4$
Oxidation
KMnO$_4$/alumina
Carbon–carbon bond formations
Glaser coupling
CuCl$_2$·2Py
Knoevenagel condensation
Me$_3$N as a catalyst
+ H$_2$O
Diels–Alder reaction
Rearrangements
Benzylic acid rearrangement
KOH
Beckmann rearrangement
FeCl$_3$, heat

FIGURE 7.6 Chemical equations for selected examples of the solid-state reactions from Table 7.1. (Tanaka, K.: *Solvent-Free Organic Synthesis*. 2009. Copyright Wiley-VCH Verlag GmbH & Co. KGaA. Reproduced with permission.)

REVIEW QUESTIONS

7.1 Write the structures of all the stereoisomers of truxillic acid. Recall that only one is obtained by the photochemical reaction.

7.2 Calculate the atom economy for the reaction shown in Figure 7.3.

7.3 a. Was there any waste in the reaction from the previous question?

   b. If yes, how much waste, and should this waste be recycled?

7.4 Rearrangement reactions in general occur with 100% atom economy, because all atoms of the starting material end up in the product. Many such reactions occur in the solid state. We bring up the high specificity of these reactions as an additional desirable factor. How are atom economy and reaction specificity related?

7.5 In the context of the previous question, examine Beckmann rearrangement, which is shown in Figure 7.6. Why does this reaction give only one product?

ANSWERS TO REVIEW QUESTIONS

7.1 There are five isomers of truxillic acid. They are shown in Figure 7.7.

7.2 $[147/(148 + 60)] \times 100 = 71\%]$

7.3 a. Yes, there is waste

   b. It is calculated as 29% (100% − 71%). A balance of atoms in the equation shows that a "CO_2NH_3" is missing on the product side. This could be $HO-C(=O)NH_2$, which could fall apart to CO_2 and NH_3. If so, the recycling would not be recommended.

7.4 If a rearrangement gives two (or more) different isomers, the atom economy is still 100%. However, if only one isomer is desired, the atom economy for that isomer drops. If we get a single specific product, which is desired, then the atom economy is truly 100%.

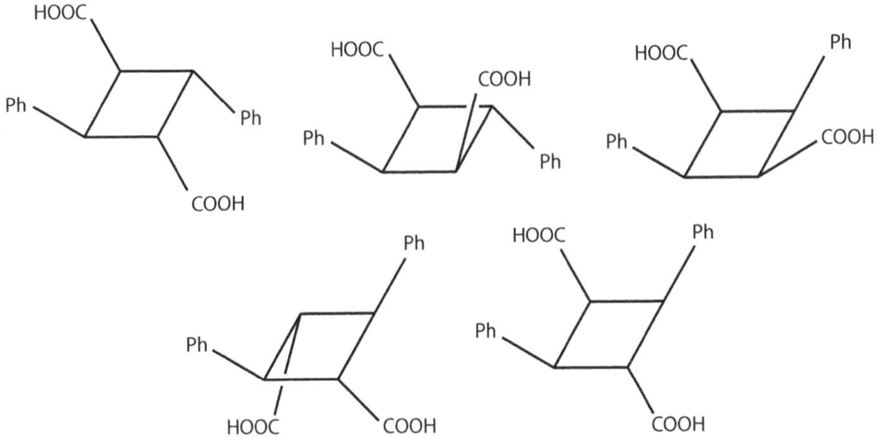

FIGURE 7.7 Isomers of truxillic acid.

7.5 This is the consequence of the reaction mechanism. The oxime that is obtained via a reaction of the corresponding ketone with hydroxylamine, HO–NH$_2$, gives the *E* isomer, which is shown in Figure 7.6. The functional group that migrates during rearrangement is in the *anti* position to the –OH of the oxime. This results in a single product, which is shown in Figure 7.6. For the mechanism of this rearrangement, consult your basic organic textbook or the article by Southam (1976). This example shows that one needs to know reaction mechanisms to fully understand why some products are favored and why some other are not.

REFERENCES

Cave, G. W. V., Rastow, C. L., and Scott, J. L. (2001). Recent advances in solventless organic reactions: Towards benign synthesis with remarkable versatility, *Chem. Commun.*, 2159–2169.

Doxsee, K. M. and Hutchinson, J. E. (2004). *Green Organic Chemistry, Strategies, Tools, and Laboratory Experiments*, Thomson Brooks/Cole, Toronto, Canada, pp. 206–210.

Kaupp, G., Schmeyers, J., and Boy, J. (2000). Quantitative solid-state reactions of amines with carbonyl compounds and isothiocyanates, *Tetrahedron*, 56, 6899–6911.

Leadbeater, N. E. and McGowan, C. B. (2013). *Laboratory Experiments Using Microwave Heating*, CRC Press, Boca Raton, FL.

Musiol, R., Tyman-Szram, B., and Polanski, J. (2006). Microwave-assisted heterocyclic chemistry for the undergraduate organic laboratory, *J. Chem. Ed.*, 83, 632–633.

Nakamatsu, S., Toyota, S., Jones, W., and Toda, F. (2005). The important role of solvent vapor in an organic solid state reaction, *Chem. Commun.*, 3808–3810.

Rastogi, R. P., Singh, N. B., and Singh, R. P. (1977). Organic solid-state reactions, *J. Solid State Chem.*, 20, 191–200.

Rothenberg, G., Downie, A. P., Raston, C. L., and Scott, J. L. (2001). Understanding solid/solid organic reactions, *J. Am. Chem. Soc.*, 123, 8701–8708.

Southam, R. M. (1976). Studies for the organic qual lab, *J. Chem. Ed.*, 53, 34–36.

Tanaka, K. (2009). *Solvent-Free Organic Synthesis*, Wiley-VCH, Weinheim, Germany.

Tanaka, K. and Toda, F. (2000a). Solvent-free organic synthesis, *Chem. Rev.*, 100, 1026.

Tanaka, K. and Toda, F. (2000b). Solvent-free organic synthesis, *Chem. Rev.*, 100, 1025–1074.

Thomas, J. M. (1979). Organic reactions in the solid state: Accident and design, *Pure Appl. Chem.*, 51, 1065–1082.

Toda, F. (1993). Solid state organic reactions, *Synlett*, 5, 303–312.

Toda, F. (1995). Solid state organic chemistry: Efficient reactions, remarkable yields, and stereoselectivity, *Acc. Chem. Res.*, 28, 480–486.

Toda, F., Takumi, H., and Akehi, M. (1990a). Efficient solid-state reactions of alcohols: Dehydration, rearrangement, and substitution, *J. Chem. Soc. Chem. Commun.*, Issue 18, 1270–1271. DOI: 10.1039/C39900001270.

Toda, F., Tanaka, K., and Hamai, K. (1990b). Aldol condensations in the absence of solvent: Acceleration of the reaction and enhancement of the stereoselectivity, *J. Chem. Soc. Perkin. Trans.*, 1, 3207–3209.

8 Applications of Green Chemistry Principles in Engineering
Introduction to Sustainability

An important point, often overlooked, is that the concept of waste is human. Green engineering focuses on how to achieve sustainability through science and technology.

Paul T. Anastas and Julie B. Zimmerman (2003)

LEARNING OBJECTIVES

This chapter is the first one in the series of five chapters that are dedicated to applications of green chemistry. In this chapter, we give a foundation of green engineering. Specific objectives are as follows.

Learning Objectives	Section Numbers
Become familiar with the basics of engineering and its main branches	8.1
Learn about the 12 principles of green engineering	8.2
	8.2.1
	8.2.2
	8.2.3
	8.2.4
	8.2.5
	8.2.6
	8.2.7
	8.2.8
	8.2.9
	8.2.10
	8.2.11
	8.2.12
Compare principles of green chemistry and green engineering	8.3
Learn essentials of sustainability and how it relates to green engineering	8.4
	8.4.1
	8.4.2
	8.4.3
Learn about types of thinking which are useful in green engineering	8.5

8.1   ENGINEERING AND ITS BRANCHES

Because engineering is not traditionally covered in the beginning chemistry classes, we provide a basic background on this topic. First, we define engineering and then chemical engineering. We further narrow down the engineering field to reflect the emphasis of this textbook on organic chemistry.

A common definition of engineering, as used in everyday life and as found in the dictionaries, is given broadly as follows:

> Engineering is a branch of science which is concerned with practical applications of the knowledge of pure sciences, such as physics or chemistry, which result in designing and constructing engines, bridges, buildings, ships, chemical plants, mines, etc. Engineering sometimes includes art, such as in its architectural applications. (Wikipedia)

A more detailed definition is needed for our use. A good coverage can be found in Wikipedia.

The word "engineering" comes from Latin *ingenium*, which means "cleverness," and *ingeniare*, which means "to devise." A broader definition of engineering includes application of scientific, economic, social, and practical knowledge, with the objectives to invent, design, build, and improve structures, machines, devices, systems, materials, and processes. This particular definition resonates well with many of the green chemistry objectives in general. For example, we notice the invention factor, which is at the core of the green chemistry movement.

This definition is still quite broad, because it encompasses all branches of engineering. Typically, engineering is divided into four main branches: chemical, civil, electrical, and mechanical.

Chemical engineering is most relevant for the objectives of this textbook. It covers chemical processes on a commercial scale. Civil engineering is concerned with design and construction of roads, railways, bridges, dams, and buildings, for example. Electrical engineering is focused on electrical and electronic systems, telecommunication, generators and motors, and similar. Mechanical engineering deals with mechanical systems, such as power, energy, aerospace, and transportation, as some examples.

There are numerous subspecialties of engineering, such as automotive, computer, architectural, agricultural, biomedical, petroleum, textile, and nuclear, among many others. Many of these fields and subfields are interconnected.

For our purposes, we shall focus on chemical engineering, and then we shall emphasize organic chemistry within a broad class of chemical manufacturing processes. Still, many organic chemical processes may cross boundaries with other engineering branches.

8.2   GREEN ENGINEERING AND ITS 12 PRINCIPLES

We now present the 12 principles of green engineering and discuss each of the principles, with the emphasis on the organic chemistry applications. We show these principles in Table 8.1, in a simplified form, which is an adaptation of several

TABLE 8.1
A Brief Description of the 12 Principles of Green Engineering

1. *Inherent rather than circumstantial*: Design the inputs and outputs of materials and energy to be as inherently nonhazardous as possible.
2. *Prevention instead of treatment*: It is better to prevent waste than to treat it or clean it up after it is formed.
3. *Design for separation*: Design separation and purification operations to minimize energy consumption and materials use.
4. *Maximize efficiency*: Design products, processes, and systems to maximize efficiency of mass, energy, space, and time.
5. *Output-pulled versus input-pushed*: Design products, processes, and systems to become "output-pulled" rather than "input-pushed," by an appropriate use of energy and materials.
6. *Conserve complexity*: Consider conservation of embedded entropy and complexity while making choices on recycling, reuse, or beneficial disposition of materials.
7. *Durability rather than immortality*: Target durability rather than immortality while designing materials and products.
8. *Meet need, minimize excess*: Design only to meet need. Design for unnecessary capacity or capability should be considered a design flaw. This includes "one-size-fits-all" solutions.
9. *Minimize material diversity*: Design multicomponent products in such a way to minimize material diversity, to promote disassembly (for reuse or recycling), and to value retention.
10. *Integrate material and energy flows*: Design products, processes, and systems in such a way to include integration and interconnectivity with available energy and material flows.
11. *Design for commercial afterlife*: Design products, processes, and systems for performance in a commercial "afterlife."
12. *Renewable rather than depleting*: Material and energy inputs should be renewable rather than depleting.

Sources: Anastas, P. T. and Zimmerman, J. B., *Environ. Sci. Technol.*, 37, 94A–101A, 2003; McDonough, W. et al., *Environ. Sci. Technol.*, 37, 434A–441A, 2003; Tang, S. et al., *Green Chem.*, 10, 268–269, 2008.

---

descriptions from the literature (e.g., Anastas and Zimmerman, 2003; McDonough et al., 2003; Tang et al., 2008). In Sections 8.2.1 through 8.2.12 we explain these 12 principles and provide examples for them. We also clarify selected terms that the readers may not be familiar with.

8.2.1 PRINCIPLE 1: INHERENT RATHER THAN CIRCUMSTANTIAL

Engineers should first evaluate the inherent (intrinsic) characteristics of the materials they intend to use. If the materials exhibit hazardous characteristic, this will become a serious drawback of the design. Hazardous materials can cause harm to life and the environment. In addition, their handling, removal, and disposal may be extremely costly and risky, and it would require time, material, special equipment, and energy resources. This is not an economically or environmentally sustainable approach. Instead, a prudent approach would be to search for other materials that are inherently as nonhazardous to life and environment as possible. In addition, designers need to ensure also that all energy inputs and outputs are inherently as nonhazardous as possible. This is especially important for large-scale manufacturing processes.

Dr. David Constable provides details about this principle (the American Chemical Society [ACS] website on green engineering, principle 1). The present-day energy sources are typically electricity and steam. These energy sources are based primarily on the fossil fuels, which are not renewable. The choice of energy also matters in terms of placing hazardous materials into the environment. One can think about gases that are generated by burning coal as one example. To minimize the output of toxic materials, engineers need to look at the type of energy that is used, in addition to the chemical reactions. Engineers need to choose intentionally both materials and energy, so as to give as nonhazardous an outcome as possible.

8.2.2 PRINCIPLE 2: PREVENTION INSTEAD OF TREATMENT

This principle states that it is better to prevent waste than to treat it or clean it up after it is formed. We recognize this principle as principle 1 of green chemistry, which we have discussed in Chapter 1. Here, we show more goals and applications that are characteristic for green engineering.

Dr. Martin Abraham (ACS website on green engineering, principle 2) brings up several specific points on how to prevent waste. First, one should make only the amount that is needed, and not an excess. The latter can be quite costly, because one pays first for the extra starting materials that are not needed, and then pays later for their safe disposal. In another example, Dr. Abraham emphasizes the need to achieve selectivity in chemical reactions. If a reaction can give two different products, where only one is desired, it is often possible to achieve selectivity in favor of the desired product. This often can be done by manipulating the temperature at which the reaction is performed.

To enforce this important point, we briefly review one of the classical reactions whose outcome can be modulated by temperature, and which is covered in every beginning organic chemistry textbook (e.g., Solomons et al., 2014a). The reaction is an addition reaction, which gives the 1,2-addition product as the main product at a low temperature, whereas at the higher temperature, the 1,4-addition product predominates. This is depicted in Figure 8.1.

At a low temperature (−80°C), the major product, which is obtained in 80% yield, is formed by a 1,2-addition of HBr. The minor product, which is the result of 1,4-addition of HBr, is obtained in a 20% yield. At the higher temperature (40°C), the same products are obtained, but the yields are reversed. Further, upon warming of the mixture of 1,2- and 1,4-products that are obtained at −80°C to the temperature of 40°C, the products are equilibrated to give the same mixture as originally obtained at 40°C. If the desired product is 1,2-adduct, one should run the reaction at low temperature. Should one desire the 1,4-adduct, one should run the reaction at a higher temperature.

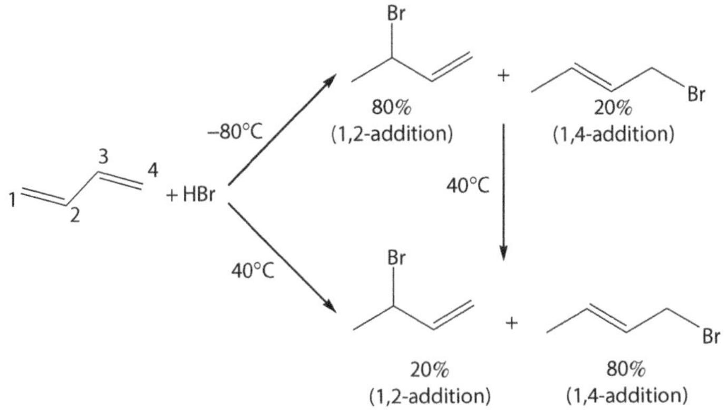

FIGURE 8.1 Ionic addition of HBr to 1,3-butadiene at different temperatures. Numbers 1–4 correspond to the position of the carbon atom in the chain, as shown on the structure on the left. (Solomons, T. W. G. et al.: *Organic Chemistry*, 11th edn. 2014a. Copyright Wiley-VCH Verlag GmbH & Co. KGaA. Reproduced with permission.)

8.2.3 PRINCIPLE 3: DESIGN FOR SEPARATION

This principle addresses the observation that separation, isolation, and purification of products generally consume most of the materials and energy requirements in manufacturing processes. Typically, separation methods require the use of toxic solvents, which then necessitate large amounts of energy to evaporate the solvents and to retrieve the products. One example on a laboratory scale that students are familiar with is the separation and purification of mixtures by column chromatography. This method requires the collection of multiple fractions as they are eluted from the column by the mobile phase, which is typically composed of volatile organic solvents. The latter may be hazardous, toxic, and/or flammable. Later, the solvents are removed from the individual fractions by distillation, which consumes energy.

Design for separation avoids the above problems because it focuses on self-separation of products. One such example is the multicomponent Passerini reaction in which three starting materials are placed in an aqueous medium, and a single product separates as a precipitate, in a quantitative yield (see Section 4.2.2). This reaction is one of the "on-water" reactions we covered in Chapter 4. Other on-water reactions, such as Diels–Alder reaction (Section 4.2.1), also lend themselves to an easy self-separation process, because they give water-insoluble products. All that is required is to filter the solid product from the aqueous medium. It should be pointed out that in the past, classical Passerini and Diels–Alder reactions, as well as many of the on-water reactions, have traditionally been performed in organic solvents, which then required separation of products from them. Only after these reactions were "greened" by performing them on water, the design for separation was realized. However, not all chemical reactions can be run on water, and for many reactions the use and distillation of solvents are still common. For such cases, green engineering offers other successful strategies (Drs. Matthew J. Realff and David Wang, Sr., ACS website on green engineering, principle 3).

For example, we give here a couple of such strategies. Distillation can be made more efficient by designing the fractional distillation towers such that the heat of condensation that is released is reused. Distillation may be avoided by using alternative separation techniques, such as the reverse osmosis membrane separation for desalination of saltwater to freshwater.

8.2.4   Principle 4: Maximize Efficiency

This principle states that products, processes, and systems should be designed to maximize efficiency of use of mass, energy, space, and time. Specifics are provided, for example, by Dr. Michael A. Gonzalez (ACS website on green engineering, principle 4).

To achieve mass efficiency, the reactions should be designed in such a way so as to use as much of the reactants as possible. Desired properties of the reactions include high conversions and selectivity to give the desired products and to give a minimal amount of by-products.

Energy efficiency is achieved by employing room temperature and pressure, when possible. This eliminates the need for expensive cooling or heating. Minimization of movement of the materials also saves energy, for example, for pumping the materials for transport. Further, if the design minimizes separation and purification steps, this also saves energy that would be otherwise needed for the distillation of solvents.

Saving space is important, because it saves on cost and reduces hazards. For example, constructing large reaction vessels is expensive in terms of cost and energy. Large volumes of the produced materials may pose unacceptable hazards. The smaller the better, providing that it meets the need.

Time efficiency strives to perform chemical reactions as fast as possible, to liberate the reactor for use of other reactions. Thus, reactions should be catalytic, if possible.

All these different aspects of efficiency—mass, energy, space, and time—are interconnected to a degree.

8.2.5   Principle 5: Output-Pulled versus Input-Pushed

This is a concept related to chemical equilibrium and states that products, processes, and systems should be "output-pulled" rather than "input-pushed" through the use of energy and materials.

Anastas and Zimmerman (2003) provide a clear example of this principle by invoking the well-known Le Châtelier's principle. First, we review this important principle in its general form. Then we apply it to an organic reaction, because the emphasis of this book is on the organic chemistry. Finally, we show its application to green engineering.

In Figure 8.2, we show a general equation for a reversible reaction, which would be a subject to Le Châtelier's principle.

aA + bB $\rightleftharpoons$ cC + dD

$$K = \frac{[C]_{eq}^{c}[D]_{eq}^{d}}{[A]_{eq}^{a}[B]_{eq}^{b}}$$

$$Q = \frac{[C]^{c}[D]^{d}}{[A]^{a}[B]^{b}}$$

FIGURE 8.2 An example of a generalized reversible reaction, which is the subject to the Le Châtelier's principle. K is the equilibrium constant. Q is the reaction quotient, whether or not the system is at equilibrium. $Q = K$ when the system is at equilibrium. a–d are the stoichiometric coefficients for the reacting species A–D. (From Brown, L. S. and Holme, T. A., *Chemistry for Engineering Students*, Thomson Brooks/Cole, Belmont, CA, 2006.)

Le Châtelier's principle summarizes the ways in which a system at equilibrium responds to changes. When stress is applied to such a system, the system responds by reestablishing the equilibrium in order to relieve, reduce, or offset the applied stress.

Common ways to introduce stress on the system at equilibrium are changes in concentrations and temperature. For gas-phase reactions, changes in pressure/volume also cause stress, providing that the number of moles of gas differs between the reactants and products. We summarize the stress and the system's response in Tables 8.2 through 8.4, which are based mostly on material from Brown and Holme (2006) and Glanville (2001). We discuss the relevance to principle 5 for each type of stress and the system's response.

The reversible reaction from Figure 8.2 at equilibrium would give us a mixture of starting materials and products. Typically, only one product is desired. How can we achieve getting only the desired product, instead of a mixture which is the consequence of the equilibrium? Inspection of the system responses from Table 8.2 shows us the options on how to achieve this goal and helps us make a green choice of

---

TABLE 8.2

Response of the System at Equilibrium from Figure 8.2 to Changes in Concentration and by Addition of Specific Components

Concentration Change	Resulting Changes in Q and K	Shift in Equilibrium	Consequence
Increase in [products]	$Q > K$	To the left	More reactants are formed
Decrease in [products]	$Q < K$	To the right	More products are formed
Increase in [reactants]	$Q < K$	To the right	More products are formed
Decrease in [reactants]	$Q > K$	To the left	More reactants are formed

Component Added	Consequence as Equilibrium Is Established		
[A]	[B] Decreases	[C] Increases	[D] Increases
[B]	[A] Decreases	[C] Increases	[D] Increases
[C]	[D] Decreases	[A] Increases	[B] Increases
[D]	[C] Decreases	[A] Increases	[B] Increases

Sources: Brown, L. S. and Holme, T. A., *Chemistry for Engineering Students*, Thomson Brooks/Cole, Belmont, CA, 2006; Glanville, J. O., *General Chemistry for Engineers*, preliminary edn., Prentice Hall, Upper Saddle River, NJ, pp. 310–317, 2001. With permission.

output-pulled versus input-pushed. One can shift the equilibrium toward the products by two means: (1) by decreasing concentrations of products, for example, by removing them from the reaction mixture, or (2) by increasing concentrations of reactants, for example, by adding extra reactants. The choice 1 is output-pulled and is green, whereas the choice 2 is input-pushed and is not green. One can easily see that in choice 2 at the end of the process, we would be stuck with the extra unused reactants. The output-pulled option can also be used to select only one of the products. We show an example in Figure 8.3, which illustrates a Fischer esterification reaction, a common esterification process, in which an alcohol and a carboxylic acid react under acidic catalysis to give an ester. This reaction is described in every organic chemistry textbook (e.g., Solomons et al., 2014b), and is often performed in the beginning organic laboratory (e.g., Nimitz, 1991; Palleros, 2000). The reaction shown in Figure 8.3 provides a fragrant ester, which students can easily recognize as fragrance of bananas. The fragrance allows for an early detection of reaction success. Esters with other fruity fragrances, such as pineapple, strawberry, or oranges, are also often synthesized by the Fischer reaction (Palleros, 2000).

In the reaction shown in Figure 8.3, the desired product is isoamyl acetate. The equilibrium can be shifted toward this product by removing water from the reaction mixture. This can be done by using concentrated H_2SO_4 beyond the catalytic amount, typically in a stoichiometric quantity. Other ways exist to remove water. Examples include use of molecular sieves or a specialized distillation technique using a Dean–Stark trap (see the following text).

Molecular sieves are porous materials of specified pore sizes. These sieves absorb small molecules, such as water, that can fit into the pores, whereas larger molecules stay outside the pores. Thus, molecular sieves can act as desiccants. Examples of molecular sieves include various zeolites, which are microporous aluminosilicate minerals.

In the distillation that uses Dean–Stark trap, water co-distills with the organic material and falls on the bottom of the trap, whereas a less dense organic layer collects on the top of the trap, and then flows back to the distilling flask (Fessenden et al., 2001).

These methods of shifting the equilibrium by removing water are output-pulled green methods. In some esterification experiments, an excess of carboxylic acid (Palleros, 2000) is used to shift the equilibrium toward the products. This would represent an input-pushed nongreen method, because one would have unreacted carboxylic

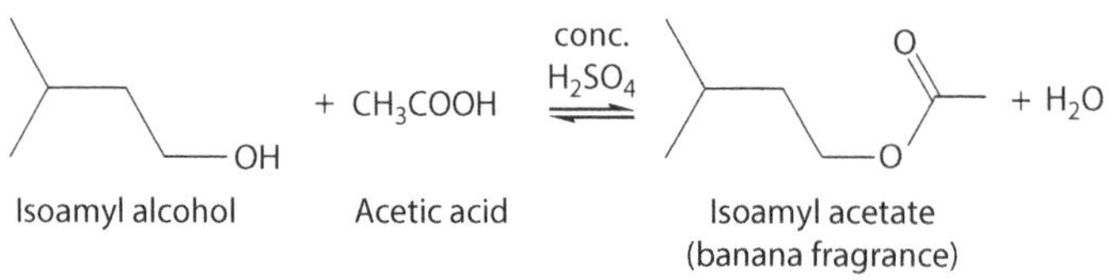

FIGURE 8.3   An example of a Fischer esterification, in which isoamyl acetate (banana fragrance) is prepared from isoamyl alcohol and acetic acid, with concentrated H_2SO_4 as a catalyst. (From Nimitz, J., *Experiments in Organic Chemistry*, Prentice Hall, Englewood Cliffs, NJ, pp. 360–361, 1991; Palleros, D. R.: *Experimental Organic Chemistry* 2000. Copyright Wiley-VCH Verlag GmbH & Co. KGaA. Reproduced with permission.)

acid left, which would need to be removed at the end. Still, one has to evaluate the entire process of shifting the equilibrium, because not all output-pulled methods are necessarily green. If one has to conduct additional steps to make the output-pulled method work, and if such steps are not green, this has to be taken into account. The above-mentioned extra distillation step to remove water is inherently not green, but may be made more efficient, and thus, the overall process is greener, by reusing the heat of condensation that is released. This would be an implementation of principle 3.

Output-pulled design can also be achieved via temperature control of reversible reactions. One can consider exothermic and endothermic reactions in a way in which one imagines that heat itself behaves as a chemical reactant or product. Chemical equations that include heat are termed thermochemical (Brown and Holme, 2006). Such equations allow us to predict the response of a system with the change in temperature in the same way as we have done previously for the change in concentration. An important difference, however, is that the temperature change alters the value of the equilibrium constant. Table 8.3 shows the response of a system in equilibrium upon a change in temperature (Brown and Holme, 2006). A more rigorous treatment of the response of equilibrium to temperature can be found in Atkins and de Paola (2006).

For the green output-pulled design, one would need to decrease the temperature for an exothermic reaction, but to increase it for an endothermic reaction.

Finally, changes in pressure can affect the equilibria of gas reactions (e.g., Glanville, 2001; Brown and Holme, 2006; for a more rigorous treatment, see Atkins and De Paula, 2006). The effect is observed only if the number of moles of gas differs between reactants and products. If an inert gas is added to the system, it will not affect the equilibrium, because it does not change the partial pressure of the gasses in the reaction. Figure 8.4 shows two gas reactions whose equilibria are affected by changes in pressure, whereas Table 8.4 summarizes the effect of changing pressure on equilibria of these reactions (Glanville, 2001; Brown and Holme, 2006).

For the output-pulled green design, one needs to decrease the pressure if the total number of the moles of products is larger than that of the reactants, such as in

---

TABLE 8.3

Response of a System at Equilibrium upon Temperature Change for Reactions That Are Exothermic (Reactants $\rightleftharpoons$ Products + Heat) and Endothermic (Reactants + Heat $\rightleftharpoons$ Products)

Type of Reaction	Temperature Change	Shift in Equilibrium	Consequence
Exothermic (heat is a "product")	Increase ("adding" heat)	To the left	More reactants are formed
Exothermic (heat is a "product")	Decrease ("removing" heat)	To the right	More products are formed
Endothermic (heat is a "reactant")	Increase	To the right	More products are formed
Endothermic (heat is a "reactant")	Decrease	To the left	More reactants are formed

Source: Brown, L. S. and Holme, T. A., *Chemistry for Engineering Students*, Thomson Brooks/Cole, Belmont, CA, 2006. With permission.

Reaction 1

$$NH_3 (g) + CH_4 (g) \rightleftharpoons HCN (g) + 3H_2 (g)$$

Reactants: Products:
2 moles total   4 moles total

Reaction 2

$$N_2 (g) + 3H_2 (g) \rightleftharpoons 2NH_3 (g)$$

Reactants: Products:
4 moles total   2 moles total

FIGURE 8.4 Examples of two gas reactions in which the number of moles differs between reactants and products. Reaction 1: synthesis of HCN. Reaction 2: synthesis of NH_3 (Haber–Bosch process). (From Brown, L. S. and Holme, T. A., *Chemistry for Engineering Students*, Thomson Brooks/Cole, Belmont, CA, 2006; Glanville, J. O., *General Chemistry for Engineers*, preliminary edn., Prentice Hall, Upper Saddle River, NJ, pp. 310–317, 2001.)

TABLE 8.4

Effect of Changing Pressure on Two Gas Reactions from Figure 8.4

Reaction	Pressure Change	Shift in Equilibrium	Response of System
Reaction 1: synthesis of HCN (2 moles give 4 moles)	Increase	To the left	More reactants are produced
Reaction 1: synthesis of HCN (2 moles give 4 moles)	Decrease	To the right	More products are produced
Reaction 2: synthesis of NH_3 (4 moles give 2 moles)	Increase	To the right	More product is produced
Reaction 2: synthesis of NH_3 (4 moles give 2 moles)	Decrease	To the left	More reactants are produced

Sources: Brown, L. S. and Holme, T. A., *Chemistry for Engineering Students*, Thomson Brooks/Cole, Belmont, CA, 2006; Glanville, J. O., *General Chemistry for Engineers*, preliminary edn., Prentice Hall, Upper Saddle River, NJ, pp. 310–317, 2001.

reaction 1, and to increase the pressure when the total number of moles of products is smaller than that of the reactants, as in reaction 2.

It should be noted that the equilibrium constant for a reaction is not affected by the presence of a catalyst or an enzyme, which acts as a biological catalyst. Catalysts increase the rate by which the equilibrium is established, but do not affect its position (Atkins and de Paula, 2006).

8.2.6 PRINCIPLE 6: CONSERVE COMPLEXITY

This principle states that embedded entropy and complexity must be viewed as an investment while making choices on recycling, reuse, or beneficial disposition of materials.

Complexity is related to the number of atoms and functional groups in a molecule used to produce the molecule or item. It is also a function of expenditure of materials, energy, and time. Thus, chemicals and products in general have a certain amount of complexity that is built into them. Many complex materials have high economic value. It would make no business sense to take a material with high economic value and convert it into one with a lower value (Dr. Michael A. Gonzalez, the ACS website on green engineering, principle 6). Highly complex materials should be reused, rather than recycled. In contrast, materials with minimal complexity should be recycled or disposed of.

A decision to reuse, recycle, or dispose of products needs to be made based on the evaluation of multiple factors that are involved in these processes. Anastas and Zimmerman (2003) bring up the example of a brown paper bag, which is made of a highly complex natural material. However, the complexity of this material does not warrant the time, energy, and cost for collection, sorting, processing, remanufacturing, and redistribution of the bag for its reuse. In another example, they discuss silicon computer chips. These chips have a significant amount of complexity built into them. Such complexity represents most of the economic value of this product. Recycling a silicon chip in order to recover the value of the starting materials makes no economic sense, but reusing the chip, which preserves its complexity, does. In conclusion, even though we strive to preserve complexity, we must evaluate such preservation from an economic point of view.

8.2.7 PRINCIPLE 7: DURABILITY RATHER THAN IMMORTALITY

This principle states that products should be designed to be durable rather than immortal. "Immortal" products are those that last well beyond their useful commercial life and that persist in the environment because they are not biodegradable. Such products may even have a short or single use. Immortality of products should be avoided. Instead, products need to be designed for useful lifetime and a capability for biodegradation. A good example of a successful design for durability but not immortality is that of a starch-based packing material. Such material is made from natural, nontoxic, biodegradable, and sustainable sources. At the end of its use, the so-called product's end of life, this product can be placed in compost piles. It is water soluble and is easily biodegradable. It is competitive with polystyrene packing, which is nonbiodegradable and has been shown to be persistent in the environment (Anastas and Zimmerman, 2003).

8.2.8 PRINCIPLE 8: MEET NEED, MINIMIZE EXCESS

Based on this principle, the design of materials, processes, and systems should be optimized for the anticipated need. A design that includes added capacity based on unrealistic assumptions or "worst-case" scenarios has negative and costly consequences. Making extra products that are not needed incures the cost of unnecessary materials and energy, the cost in time and salaries for the workers, the cost for storage, the burden on the environment in terms of produced waste, and the cost of disposal and recycling of waste, among others. To avoid the problem of overproduction and the associated cost and other negative consequences, the green alternative to target specific needs and demands of the end users should be implemented.

8.2.9   PRINCIPLE 9: MINIMIZE MATERIAL DIVERSITY

This principle seeks to minimize material diversity of products, namely, to make products from as few types of materials as possible. Material diversity is taken into account when considering options for the end of the useful life of the product, such as reuse or recycling. A good description of this principle is provided by Anastas and Zimmerman (2003).

Many products are built of multiple components, and such components may be composed of a variety of materials. For example, cars have components made of metal, glass, plastic, paint, and so on. Individual plastics typically contain additional materials, such as thermal stabilizers, plasticizers, and flame retardants. At the end of the useful life of the product, decisions about disassembly for reuse of parts and recycling need to be made. Such decisions ultimately depend on material diversity, because different materials have different requirements and potential for green reuse and recycling. Lowering the material diversity in the product and choosing or designing material with good potential for reuse and/or recyclability enable the implementation of the end-of-useful life options.

8.2.10   PRINCIPLE 10: INTEGRATE LOCAL MATERIAL AND ENERGY FLOWS

This principle states that optimization of mass and energy requirements for a given process is best accomplished via integration of heat exchange networks (HENs) and mass exchange networks (MENs).

As an example, let us consider the existing framework of energy and material flow at a local chemical production facility. The application of principle 10 can be understood as an integration of HENs and MENs (Dr. Concepcion Jiménez-González, ACS website on principles of green engineering, principle 10). HEN uses hot streams to heat cold streams, for example. If one production unit generates heat, and another one requires heating, the two should be integrated.

MENs are analogous to HENs, but instead of exchanging energy, they exchange mass. For example, we may need to separate unreacted material from the product, both to purify the product and to recover the unreacted material so that we can use it again in the reaction. One option is to use additional material, such as a solvent, to remove the unreacted material and to send it back to the reactor. However, such an option is costly, and the solvent may be toxic or otherwise hazardous. A better option would be to implement MEN, namely, to use streams with low concentrations of a given material ("lean" streams) to separate and recover material from streams that have high concentrations of the material in question ("rich" streams). We should use as much of the lean stream as possible to recover the material, before we resort to using additional materials, such as solvents.

Although HENs and MENs appear to be quite simple and straightforward in principle, it takes a lot of expertise and creativity to implement them.

8.2.11   PRINCIPLE 11: DESIGN FOR COMMERCIAL "AFTERLIFE"

This principle states that design of a product should include the commercial "after-life." This means that the product at the end of its commercial life should still has value, which can be extracted from it, if the product is properly designed.

For example, many products reach their commercial end of life because they become technologically or stylistically obsolete. They may still have perfectly good and valuable working components. The latter can be recovered for reuse, if it is economically feasible. Valuable components can be refurbished and sold, or may be disassembled to extract other valuable parts. Various examples are given in the work of Anastas and Zimmerman (2003). A good example is that approximately 90% of Xerox equipment is designed for remanufacturing.

8.2.12   PRINCIPLE 12: RENEWABLE RATHER THAN DEPLETING

This principle states that the material and energy inputs should be renewable rather than depleting. The ultimate goal is sustainability. For example, fossil fuels, such as petroleum, natural gas, and coal, are not renewable and their continuous use is not sustainable. In contrast, solar, wind, and hydro energy sources that generate electricity are renewable and are sustainable.

In terms of materials, bio-based plastic is one example of use of a renewable feedstock. It is important to go through all the steps of the process by which a product is obtained, before one decides on sustainability of the product. One should not automatically assume that bio-based products are always more sustainable than the identical products that are manmade. Sometimes, extracting the product from natural sustainable resources, such as plants, may be extremely costly in terms of energy and auxiliary materials, and a green synthetic organic path may be preferred.

8.3   COMPARISON OF PRINCIPLES OF GREEN CHEMISTRY AND GREEN ENGINEERING

The 12 principles of green chemistry were covered in Chapter 2. A quick comparison between these principles and the 12 principles of green engineering reveals a substantial overlap. Examples include green chemistry principle 1 and green engineering principle 2, both of which focus on waste prevention. Green chemistry principle 7 and green engineering principle 12 both cover a need for renewable resources. Green chemistry principle 3 and green engineering principle 1 both state a goal of nonhazardous processes. Green chemistry principle 10 calls for design for degradation so that products do not persist in the environment. This goal is embedded also in green engineering principle 7, which calls for durability rather than immortality. The "green" aspects, such as sustainability and design for biodegradation, are common themes for both green chemistry and green engineering.

They differ in that green engineering emphasizes the practical requirements for sustainability, renewability of resources, end of life and afterlife of products, durability versus immortality, material diversity, efficiency, conservation of complexity, and other individual factors which are described in its principles.

8.4   SUSTAINABILITY AS RELATED TO GREEN ENGINEERING

Sustainability is a broad term that denotes meeting the needs of the present without compromising the ability of future generations to meet their own needs. Sustainability has economic, social, and environmental aspects (García-Serna et al., 2007).

Sustainability goals are embedded in various disciplines. We have seen this already through the analysis of the principles of green engineering, and also green chemistry. Sustainability is an integral and critical part of these principles, and of the entire green movement. This is especially noted in green engineering, which is more oriented toward applications.

In this section, we briefly describe three common approaches to sustainability, aspects of which are put into practice in green engineering. They are the natural step (TNS), biomimicry, and cradle to cradle (C2C).

8.4.1 THE NATURAL STEP

In the TNS approach to sustainability, the economic profitability and environmental and social accountability are weighted equally (García-Serna et al., 2007). TNS seeks a sustainable capital, which consists of natural, human, social, manufacturing, and financial capital. Its goals are (1) to avoid a systematic increase in concentrations of substances in ecosphere, such as those that are extracted from the Earth's crust or produced by society; (2) to prevent systematic deterioration of the physical basis for the productivity and diversity of nature; and (3) to implement a fair and efficient use of resources in meeting basic human needs. Goal 1 is in response to the pollution of ecosphere by human activities, goal 2 is in response to human interference with nature, and goal 3 is related to wasting resources. These are also goals of green engineering, but in a more narrow and specific way.

8.4.2 BIOMIMICRY

Biomimicry approach uses nature as a model for sustainability and green engineering. In this approach, one observes, seeks to understand, and finally mimics natural solutions by inventing similar but manmade designs (Benyus, 2002).

Examples of observations of these natural solutions to engineering problems include capturing solar energy by leaves or purple bacteria, making strong fibers such as spider silk, making strong adhesives by mussels, and analyzing sustainability properties exhibited by the mature ecosystems, among many others.

A particularly instructive example of sustainability is that of redwood forests' mature ecosystems. Organisms in such systems use the output of other organisms as a resource, gather, and use energy efficiently; use materials sparingly; optimize rather than maximize; do not deplete resources; remain in balance with the biosphere; and utilize local materials (Benyus, 2002).

Especially inspiring is the way these ecosystems use the output of one species as a resource for the second. This closes the cycle of transformations of organic materials within the ecosystem. In human societies, often there is a linear path of products to waste, which then becomes a problem. Let us recall the citation by Anastas and Zimmerman (2003) from the very beginning of this chapter: "An important point, often overlooked, is that the concept of waste is human."

8.4.3 Cradle to Cradle

The C2C concept was developed in 2002 by William McDonough, an architect, and Michael Braungart, a chemist (McDonough and Braungart, 2002). It introduces a new paradigm to replace that of "cradle to grave," which dominates modern manufacturing and which is based on one-way flow of materials, from products to their "grave," such as a landfill. In the C2C method, materials flow back to the system, and thus form a closed loop. C2C shares with biomimicry a concept that the output of a system is a valuable resource. Waste, a human concept, needs to be thought of as a resource or "food," by analogy with the natural ecosystems. C2C designs the products and processes in such a way to ensure that traditional waste is not formed; instead, what is created is "food." This is summarized in the first tenet of C2C: "Waste equals food" (McDonough and Braungart, 2002; McDonough et al., 2003; Garciá-Serna et al., 2007).

C2C introduces two types of food, namely, biological and technical nutrients:

Biological nutrients are materials or products that are designed to return to the biological cycle. For example, textiles made from the natural fibers represent biological nutrients. They will biodegrade and will end up in soil as nutrients for various organisms.

Technical nutrients are materials or products that are designed to go back into technical cycles from which they originated. Examples include synthetic polymers that are designed in such a way to allow for repeated depolymerization and repolymerization. Thus, they will generate no waste, only "food" for the repeated polymer production.

The second tenet of C2C is also nature inspired. It is summarized as follows: "Use current solar income." Nature uses solar energy as its primary energy source, as evidenced by photosynthesis in which plants produce complex organic molecules by utilizing solar energy.

The third and final tenet of C2C is also of a biomimetic nature. It is stated as follows: "Celebrate diversity." This tenet reflects observation that diversity makes ecosystems resilient in their response to changes.

A detailed description of incorporation of C2C tenets into the principles of green engineering is provided by McDonough et al. (2003). The C2C concept has been further developed and expanded by its authors (McDonough and Braungart, 2013), in which they cover "upcycling," which is reuse of discarded objects or materials in such a way to create a product of a higher quality or value than the original. This would be the opposite of "downcycling," in which the quality of value of the product is less than the original.

8.5 TYPES OF THINKING USEFUL IN GREEN ENGINEERING: SYSTEMS AND HOLISTIC THINKING

Earlier in this book, we have surveyed various types of thinking as appropriate for green chemistry, such as deductive, inductive, critical, linear, nonlinear, lateral, vertical, and complex (Section 3.2). We have evaluated each type of thinking and have concluded that all of them are useful for green chemistry, but specifically complex thinking may be needed to cover the simultaneous consideration of all 12 principles. The same types of thinking are relevant to green engineering. However, we notice additional challenges that are posed by green engineering when networks and their integration are considered. One example was given in principle 10 of green engineering, which requires integration of HENs and MENs (Section 8.2.10). Methods that are particularly useful for considering networks are systems thinking (Meadows, 2008) and holistic thinking (Kasser, 2013), in addition to complex thinking, which was previously covered in Section 3.2. These three types of thinking are related. Also, there is a substantial overlap between them.

We present here a brief and simplified summary of the systems and holistic thinking, and refer the readers to the original references (Meadows, 2008; Kasser, 2013) for a more complete treatment. We start with some basics.

A system may be defined as an interconnected set of elements that are coherently organized in a way that achieves a purpose. Thus, a system consists of elements, interconnections, and purpose. Systems can be embedded in other systems, which further may be embedded in yet other systems. We can think about a chemical reactor as a system, which is embedded in a chemical factory, which is further embedded in an industrial park. It is easier to learn about a system's elements than to learn about the interconnections that hold the system's elements together. However, it is precisely the presence of interconnections which requires an expansion of our thinking. Although we may know a chemical reaction, once a reaction becomes a part of a system, such as a chemical reactor, or factory, or an industrial park, we find out that knowledge about the reaction or the way of thinking about the reaction does not longer suffice. We must be able to view the systems in terms of their interconnections.

Systems thinking provides us with a useful way to think about sustainability. Let us consider a presumably renewable resource consisting of a plant material, from which some important chemical, such as a drug, is extracted. If we take into account the entire system, comprising the plant material and including all other elements of the process all the way to the final extraction of the drug, we realize the inherent limitation of a natural plant-based source. If the flow of the plant extract into the system is higher than the regeneration rate of the plant, we shall exhaust our plant source, which then will not be renewable.

Systems thinking helps us to understand better the ecosystem. Ecosystems exhibit resilience, which is the ability to restore or repair itself. However, there are always limits to resilience of any system, and if we go above the threshold of the resilience, the system will break down. As green engineers and chemists, we must be aware of this and make sure that the system is maintained within its inherent limits of self-repair.

From these brief examples, we realize that our thinking type must match the problem that we are dealing with. More useful tips about thinking in general, systems thinking, and a new type of thinking, holistic thinking, are provided by Kasser (2013).

In the following text, we present different types of thinking, as described by Kasser. Then we introduce the need for the systems thinking and finally the holistic thinking.

Types of thinking can be absorptive (the ability to observe), retentive (the ability to memorize and recall), reasoning (the ability to analyze and judge), and creative (the ability to visualize, to foresee, and to generate ideas). Thinking can be top-down (analysis) or bottom-up (synthesis). Analysis consists of breaking down a complex problem into smaller ones, and then thinking about the latter. This thinking type is also known as reductionism, because it is used to reduce a complex problem to a number of smaller and simpler ones. In contrast, synthesis combines thinking about two or more phenomena, to form a more complicated question. For green chemistry and green engineering, we need all of these types of thinking, and we need to combine analysis and synthesis. As mentioned in Chapter 3, the green movement requires creative thinking, which will generate ideas, but also needs critical thinking, which will analyze, compare, and choose the best solutions.

An instructive example of analysis is given as a series of steps: (1) take apart the phenomenon that needs to be understood, (2) try to understand how the individual parts function independently, and (3) apply the understanding of the parts into an understanding of the whole. This is a reductionist way of thinking, as mentioned before, because it reduces the phenomenon to smaller parts.

Unfortunately, this type of thinking does not give good results when the phenomenon under study is a system. Instead, one needs systems thinking. Such thinking looks at a system as a whole and seeks to identify relationships, connectedness, and patterns within the system, rather than focusing on its parts. The systems thinking focuses on understanding relationships and their context in the system.

Still, there are more ways to improve our thinking about complex systems. Kasser describes the concept of "holistic thinking perspective" (HTP). A clear description of HTP is provided via nine anchor points within four different perspectives. This is presented in Table 8.5.

TABLE 8.5

Holistic Thinking Perspective

External perspectives
  1. Big picture: the context of the system
  2. Operational: what the system does

Internal perspectives
  3. Functional: what the system does and how does it do it
  4. Structural: how is the system constructed and organized

Progressive perspectives
  5. Generic: the system is perceived as an instance of a class of similar systems
  6. Continuum: the system is perceived as one of many alternatives
  7. Temporal: considers the past, present, and future of the system

Other perspectives
  8. Quantitative: considers numeric and other quantitative information associated with the system
  9. Scientific: generates hypotheses about the system

Source: Kasser, J., *Holistic Thinking, Creating Innovative Solutions to Complex Problems,* The Right Requirement, Bedfordshire, 2013. With permission.

Note: This type of thinking perspective is well suited for green engineering.

Each of the points that are presented in Table 8.5 are further discussed by Kesser. Finally, he introduces holistic thinking, which is a combination of the use of HTP and the evaluation of ideas by critical thinking.

Both Meadows (2008) and Kasser (2013) provide specific examples, case studies, and instructions on how to achieve the proper way of thinking about systems.

8.6 CONCLUSIONS

In this chapter, we have described the 12 principles of green engineering, sustainability principles, and their selected methods. We have also discussed the methods of thinking, which are well suited for the analysis of systems and networks that are characteristic in engineering. It is important to realize that the field of engineering is broader than that of chemistry. Still, green chemistry and green engineering are related in many respects. As we move into the further applications of green chemistry, we shall find it necessary to revisit some of the principles of green engineering and related sustainability. The interdisciplinary nature of green chemistry will become even more apparent, because we shall bring into play industrial chemistry, pharmaceutical chemistry, and other fields.

REVIEW QUESTIONS

8.1 In some Fisher esterification reactions, water is removed by adding concentrated H_2SO_4 above the needed catalytic amount. Is this method green? What alternatives would be better?

8.2 Show the direction of the equilibrium shift while (1) increasing and (2) decreasing the pressure for the flowing equilibrium gas reactions (Glanville, 2001; Brown and Holme, 2006).

Reactions:

a. $H_2(g) \rightleftharpoons 2H(g)$
b. $2NO_2(g) \rightleftharpoons N_2O_4(g)$
c. $2NH_3(g) + 2CH_4(g) + 3O_2(g) \rightleftharpoons 2HCN(g) + 6H_2O(g)$
d. $CO_2(g) + H_2(g) \rightleftharpoons CO(g) + H_2O(g)$
e. $2H_2(g) + O_2(g) \rightleftharpoons 2H_2O(g)$

8.3 Practice systems thinking, by first identifying the systems. An example is as follows: Is your university a system? If so, what are its interconnected parts? What is its goal?

8.4 What is wrong with the thinking that we should not bother with sustainability because the Earth will resist any adverse effects of exploiting its resources?

8.5 Compare the C2C approach with the 12 principles of green engineering. Specifically comment on the goals of tenet 2 of C2C and principle 12.

8.6 Compare the property of the mature redwoods in which they optimize but not maximize their growth with the appropriate principles of green engineering.

8.7 Give examples of products that can be described by the cradle-to-grave model.

8.8 Give examples of products that can be described by the C2C model.

ANSWERS TO REVIEW QUESTIONS

8.1 Because concentrated H_2SO_4 is quite hazardous, the method is not green. Molecular sieves may be a better choice in some cases, because they are not inherently hazardous and can be regenerated for further use.

8.2 a. 1 mole gives 2 moles: (1) to the left; (2) to the right
    b. 2 moles give 1 mole: (1) to the right; (2) to the left
    c. 7 moles give 8 moles: (1) to the left; (2) to the right
    d. 2 moles give 2 moles: no change
    e. 3 moles give 2 moles: (1) to the right; (2) to the left

8.3 Answers will vary. An example is as follows: A university is a system composed of its elements, such as students, instructors, administrators, admission officers, dorm personnel, library personnel, cleaning personnel, and health services personnel, all of which are interconnected, with the ultimate goal to provide an education to students. A university is a system with nested goals, and thus subgoals. For example, health services personnel have a goal to help students with health-related issues.

8.4 The systems that exhibit resilience, such as the Earth, all show that resilience is not unlimited, and at some point, the system will break down beyond repair if it becomes stressed too much.

8.5 Tenet 2 recommends the use of solar energy, which is also a goal of principle 12, according to which energy should be renewable.

8.6 It is principle 8, in which design should fit desired capacity or capability, and principle 10, which aims toward integration and interconnectivity of available energy and materials flows.

8.7 Answers will vary. An example is use of Styrofoam coffee cups: after a single use, they will be discarded and will end up in a landfill as their grave. Because Styrofoam is not biodegradable, they will remain in the landfill for a long time.

8.8 Answers will vary. An example is newspapers that can be recycled and remanufactured to other paper products.

REFERENCES

American Chemical Society website on green engineering, https://www.acs.org/content/acs/en/greenchemistry/what-is-green-chemistry/principles/12-principles-of-green-engineering.html, accessed on February 24, 2016. (On this website, there are special contributions by different authors describing specific principles of green engineering. We cite these authors in the text, but the website as the reference.)

Anastas, P. T. and Zimmerman, J. B. (2003). Design through the twelve principles of green engineering, *Environ. Sci. Technol.*, 37, 94A–101A. (Citations are from p. 96A and 94A, respectively.)

Atkins, P. and de Paula, J. (2006). *Atkins' Physical Chemistry*, 8th edn., W. H. Freeman and Company, New York, pp. 210–214.

Benyus, J. M. (2002). *Biomimicry: Innovation Inspired by Nature*, William Morrow/HarperCollins Publishers, New York.

Brown, L. S. and Holme, T. A. (2006). *Chemistry for Engineering Students*, Thomson Brooks/Cole, Belmont, CA.

Engineering: http://en.wikipedia.org/wiki/Engineering. Accessed on March 11, 2015.

Fessenden, R. J., Fessenden, J. S., and Feist, P. (2001). *Organic Laboratory Techniques*, 3rd edn., Brooks/Cole, Pacific Grove, CA, pp. 73–74.

García-Serna, J., Pérez-Barrigón, L., and Cocero, M. J. (2007). New trends for design towards sustainability in chemical engineering: Green engineering, *Chem. Eng. J.*, 133, 7–30.

Glanville, J. O. (2001). *General Chemistry for Engineers*, preliminary edn., Prentice Hall, Upper Saddle River, NJ, pp. 310–317.

Kasser, J. (2013). *Holistic Thinking: Creating Innovative Solutions to Complex Problems*, The Right Requirement, Bedfordshire.

McDonough, W. and Braungart, M. (2002). *Cradle to Cradle: Remaking the Way We Make Things*, North Point Press, New York.

McDonough, W. and Braungart, M. (2013). *The Upcycle: Beyond Sustainability—Designing for Abundance*, North Point Press, New York.

McDonough, W., Braungart, M., Anastas, P. T., and Zimmerman, J. B. (2003). Applying the principles of green engineering to cradle-to-cradle design, *Environ. Sci. Technol.*, 37, 434A–441A.

Meadows, D. H. (2008). *Thinking in Systems: A Primer*, Chelsea Green Publishing, White River Junction, VT.

Nimitz, J. (1991). *Experiments in Organic Chemistry*, Prentice Hall, Englewood Cliffs, NJ, pp. 360–361.

Palleros, D. R. (2000). *Experimental Organic Chemistry*, Wiley, Hoboken, NJ, pp. 473–490.

Solomons, T. W. G., Fryhle, C. B., and Snyder, S. A. (2014a). *Organic Chemistry*, 11th edn., Wiley, Hoboken, NJ, pp. 605–607.

Solomons, T. W. G., Fryhle, C. B., and Snyder, S. A. (2014b). *Organic Chemistry*, 11th edn., Wiley, Hoboken, NJ, pp. 790–791.

Tang, S., Bourne, R., Smith, R., and Poliakoff, M. (2008). The 24 principles of green engineering and green chemistry: IMPROVEMENTS PRODUCTIVELY, *Green Chem.*, 10, 268–269.

Tsoka, C., Johns, W. R., Linke, P., and Kokossis, A. (2004). Towards sustainability and green chemical engineering: Tools and technology requirements, *Green Chem.*, 6, 401–406.

9 Chemical Industry and Its Greening
An Overview of Metrics

Many companies in chemical industry have been amazed to learn that replacement of dangerous and/or toxic chemicals leads not only to safety but also to greater profit.

Mark M. Green and Harold A. Wittcoff (2003a)

LEARNING OBJECTIVES

The learning objectives for this chapter are as follows.

Learning Objectives	Section Numbers
Become familiar with the scope of chemical industry and with selected categories of common industrial products	9.1
Become familiar with products of polymer industry	9.2
Learn about different types of polymers that are responsible for polymer properties	9.3
Learn about greening of the chemical industry via specific examples	9.4
	9.4.1
	9.4.2
	9.4.3
Learn about metrics in green chemistry which are essential for greening of chemical industry	9.5

9.1 INTRODUCTION: SCOPE OF CHEMICAL INDUSTRY AND THE OBJECTIVES OF THIS CHAPTER

In this section, we first address the scope of chemical industry and describe selected categories of common industrial products (Wikipedia on chemical industry, on plastic, on polymer; Stevens, 1999; Stine, 1994; Green and Wittcoff, 2003a). This is necessary because this information is traditionally not included in the beginning organic chemistry textbooks in a systematic manner, although specific examples may be occasionally given.

Chemical industry is a branch of enterprises, such as chemical companies, that manufacture chemicals. It is an important part of the world's economy, because chemicals represent about three trillion dollars of global enterprise.

It is estimated that chemical industry produces more than 70,000 products. These products are often divided into four categories: (1) "basic or commodity chemicals," such as polymers, bulk petrochemicals, inorganic chemicals, and fertilizers; (2) "life science products," such as pharmaceuticals, vitamins, diagnostics, animal health products, and pesticides; (3) "specialty chemicals," such as adhesives and sealants, coatings, paints, pigments, inks, cleaning chemicals, and various additives; and (4) "consumer products," such as soaps, detergents, and cosmetics.

Chemical industry produces inorganic chemicals, such as salt, chlorine, caustic soda (NaOH), soda ash (Na_2CO_3), acids (such as nitric, phosphoric, and sulfuric), titanium dioxide, and hydrogen peroxide, among others. This is the oldest branch of chemical industry. We shall not cover it here, because the emphasis of this book is on organic chemistry. For the same reason, we shall not cover fertilizers, such as phosphates, ammonia, and potash (K_2CO_3).

Polymers comprise the largest output of the chemical industry, about 80%. They also create about one-third of the revenue in the basic chemicals category. Polymers are large molecules made up of many repeating subunits, which are called monomers. Monomers join by covalent bonds to form polymers in a so-called polymerization reaction. The principal raw materials for polymers are bulk petrochemicals, which are primarily made of liquefied petroleum gas, natural gas, and crude oil. These are nonrenewable sources, and are thus not green. In addition, many polymers are not biodegradable, another property that makes them not green. Due to the central role of polymers in industrial chemistry and the need for polymer chemistry to be greened, we cover it further in this chapter.

Pharmaceutical industry is another area of industrial chemistry that deserves a special focus. Pharmaceuticals are typically high-cost products. They are often complicated compounds with specific stereochemical requirements. Their synthesis is often elaborate, and it typically involves numerous steps. Much research is needed to design pharmaceutical chemical structures and to put the synthesis into practice. Purification requirements are stringent, because extensive testing is typically required to determine if the products are suitable for a particular biological application. Impurities could skew the results of the biological tests. There are multiple areas of pharmaceutical industry which require greening. Many are related to the syntheses themselves that often use toxic chemicals and solvents, and copious amounts of energy. We shall address pharmaceutical industry in Chapter 10, but will provide a foundation for much of its greening in this chapter by introducing the topic on metrics in green chemistry. This topic is universally important for all branches of chemical industry.

Another universal topic for chemical industry is a heavy use of various catalysts. Such catalysts are often based on metals and are quite toxic. We address in this chapter the subject of greening chemical catalysis, again via specific examples.

9.2 PRODUCTS OF POLYMER INDUSTRY

As we have learned in Section 9.1, a large majority of products of chemical industry are polymers. In this section, we shall learn more about various common products of polymer industry.

There are five major classifications of polymer industry: "plastics, fibers, rubber (elastomers)," "adhesives, and coatings."

"Plastics" are polymeric materials that are malleable. Thus, they can be molded into solid objects of various shapes. They often contain substances other than the main component polymers to improve performance and/or to reduce cost (Wikipedia on plastic; Stevens, 1999).

Plastics are typically divided into "commodity plastics," which are high volume and low cost, and "engineering plastics," which are lower volume and higher cost.

"Commodity plastics" are often used in disposable items, such as packaging film, but also in durable goods, such as appliances, flooring, and carpeting. "Engineering plastics" typically have mechanical properties and durability. They are used in products such as aircrafts, computers, and automobiles.

"Fibers, rubber (elastomers)," and "adhesives, and coatings" can be natural or synthetic. In this chapter, we focus on the synthetic products, because they are typically the candidates for greening.

Table 9.1 provides examples of industrial polymers and their common uses (Wikipedia on chemical industry and plastic; Stevens, 1999; Stine, 1994).

Figure 9.1 shows chemical structures of selected addition polymers from Table 9.1 and their corresponding monomers, whereas Figure 9.2 depicts selected condensation polymers from Table 9.1, their corresponding monomers, and condensation by-products. The basic mechanisms by which these types of polymers are prepared can be found in any beginning organic chemistry textbook (e.g., Solomons et al., 2014).

9.3 TYPES OF POLYMERS AFFECT POLYMER PROPERTIES

In this section, we address some chemical challenges of polymer chemistry in designing and preparing polymers with desired mechanical and other properties. Such challenges may not be covered in the basic organic chemistry classes. The inspection of Figures 9.1 and 9.2, which show the structures of the monomers and the resulting polymers, suggests that making polymers by chemical connections of monomers is straightforward. This is not the case. Also, there are numerous polymer products that have been listed in Table 9.1, which have diverse properties. How are these achieved? We offer here some information and answers. Much more can be found in the references on polymer chemistry that are listed at the end of this chapter.

Polymerization of vinyl monomers, $CH_2=CH-X$ (where X is CH_3, Cl, Ph, CN, etc.) occurs in a head-to-tail fashion. This places the X group on every other carbon atom along the polymer chain. Upon polymerization, the C–X carbon atom becomes chiral. Depending on the position of the X group with respect to the

TABLE 9.1
Examples of Industrial Polymers and Their Common Uses

Polymer (Abbreviation) [Familiar or Trade Names]	Common Uses
Polyethylene (PE); it can be a low density (LD) or a high density (HD)	Packaging film, milk bottles, housewares, coatings, wire and cable insulation, supermarket bags
Polypropylene (PP)	Appliances, automobile parts, such as bumpers, carpeting, furniture
Polyvinyl chloride (PVC)	Rigid pipes, such as plumbing pipes, siding in construction, flooring
Polystyrene (PS)	Packaging, insulation foam, food containers, housewares, toys
Polymethyl methacrylate (PMMA) [Lucite, Plexiglas]	The original hard contact lenses, fluorescent light diffusers, the basis for acrylic paints
Polytetrafluoroethylene (PTFE) [Teflon]	Heat-resistant, low-friction coatings, used in frying pans and water slide
Polyacrylonitrile [Orlon, Acrilan]	Fibers for rugs and clothing
Polyesters (PESs) such as polyethylene terephthalate (PET) [Fortrel, Mylar, Dacron]	Fibers, textiles, carpeting, tire cords, electrical insulation, plastic soda bottles
Polyamides (PAs) [Nylons]	Fibers, toothbrush bristles, fishing lines
Polycarbonates (PCs)	Compact discs, eyeglasses, security windows
Polyurethanes	Foams in furniture, packing materials for cushioning, sponges, insulation
Formaldehyde polymers: phenol formaldehyde (PF) [Bakelite] and melamine formaldehyde (MF)	PF: in electrical insulation, it is used as handles for cooking utensils, buttons, and where hard and durable plastic is required; MF: in making dinnerware, surface coating for counter tops and tables

Sources: Chemical industry: https://en.wikipedia.org/wiki/Chemical_industry, accessed on June 4, 2015; Plastic: wikipedia: https://en.wikipedia.org/wiki/Plastic, accessed on June 4, 2015; Stevens, M. P., *Polymer Chemistry: An Introduction*, 3rd edn., Oxford University Press, New York, 1999; Stine, W. R., *Applied Chemistry*, 3rd edn., D. C. Heath and Company, Lexington, MA, 1994.

carbon–carbon backbone plane, three stereoisomers can be obtained: (1) "atactic," in which the X groups are randomly arranged; (2) "isotactic," in which each chiral center has the same configuration, and all X groups are lined up on the same side of the backbone plane (e.g., all are "up"); and (3) "syndiotactic," in which the alternate chiral centers have the same configuration, and the X groups alternate between the sides of the plane (e.g., "up-down"). All three types of polymers can be synthesized. They have different mechanical properties. For example, atactic polypropylene has a consistency of chewing gum, whereas the other two types of polypropylene, which are stereoregular forms, are hard, rigid plastics (Rosen, 1993; Stevens, 1999). Stereoregularity of polypropylene depends on the type of the catalyst that is employed in its synthesis.

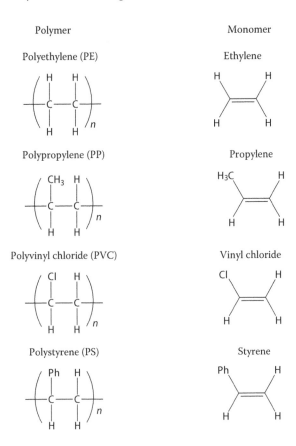

FIGURE 9.1 Chemical structures of selected addition polymers from Table 9.1 and their corresponding monomers. (From Stevens, M. P., *Polymer Chemistry: An Introduction*, 3rd edn., Oxford University Press, New York, 1999; Stine, W. R., *Applied Chemistry*, 3rd edn., D. C. Heath and Company, Lexington, MA, 1994; Solomons, T. W. G. et al.: *Organic Chemistry*, 11th edn. 2014. Copyright Wiley-VCH Verlag GmbH & Co. KGaA. Reproduced with permission.)

Polymers can be prepared from a single monomer, A, to give a homopolymer, –A–A–A–A–A–A–. If monomers A and B are polymerized together, they make copolymers. Four arrangements of copolymers are possible: (1) "alternating copolymer," –A–B–A–B–A–B–A–B–A–B–; (2) "random copolymer," –A–A–A–B–A–B–B–A–B–; (3) "block copolymer," –A–A–A–A–B–B–B–B–; and (4) "graft copolymer," which is a nonlinear block arrangement consisting of one polymer with another polymer branching from it, –A–A(B–B–B–B–B–B–B–B–)–A–A–A–A–A–. Properties of polymers depend on the structures of the monomers and types of copolymers, and can be designed to meet specifications for a particular use.

Polymers can be linear, branched, and cross-linked. These different types of polymers result in different physical and mechanical properties of polymers. Again, the latter can be designed by manipulating polymer structure.

Vinyl polymerization can be free radical, cationic, anionic, and group transfer. The stereochemical outcome of the vinyl polymerization can be influenced by the use of complex coordination catalysts, such as Ziegler–Natta catalysts, which are metal based.

Many monomers, initiators, catalysts, and solvents that are used for vinyl polymerizations are not green, typically due to their toxicity. The same is true for many starting materials for polycondensation reactions, such as isocyanate, phenol, phosgene, and formaldehyde. Thus, alternative starting materials, which are green, need

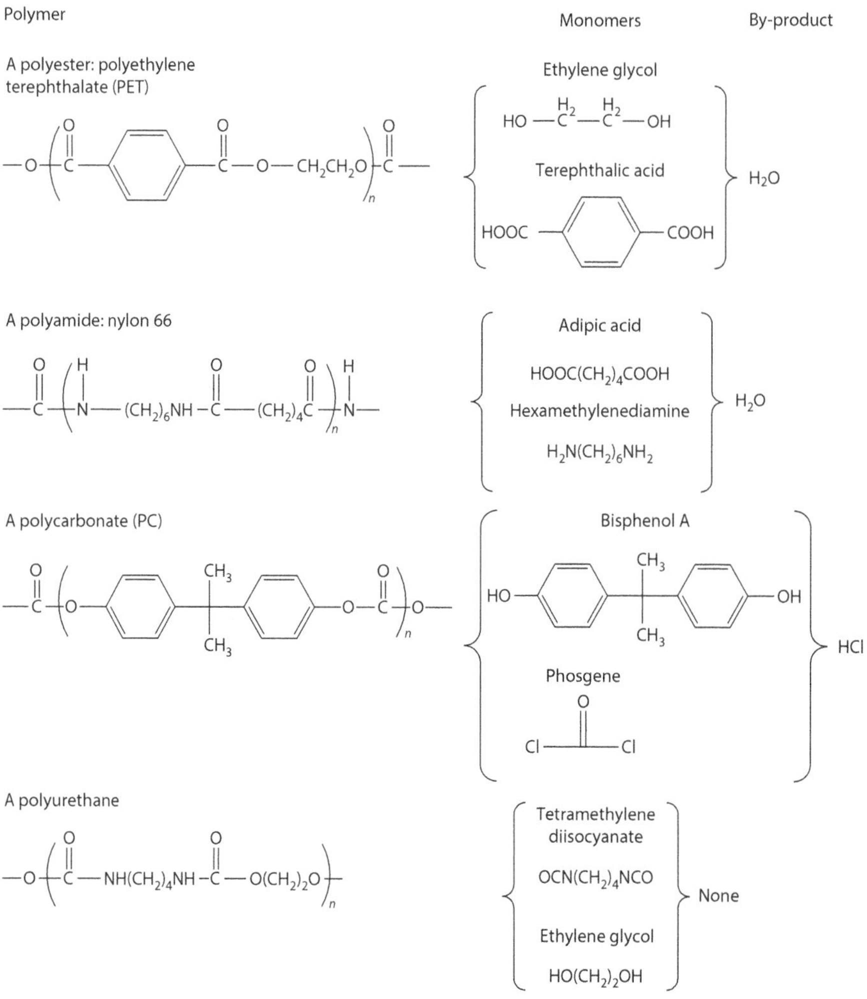

Polymer	Monomers	By-product

FIGURE 9.2   Chemical structures of selected condensation polymers from Table 9.1, their corresponding monomers, and the condensation by-products. (From Stevens, M. P., *Polymer Chemistry: An Introduction*, 3rd edn., Oxford University Press, New York, 1999; Stine, W. R., *Applied Chemistry*, 3rd edn., D. C. Heath and Company, Lexington, MA, 1994; Solomons, T. W. G. et al.: *Organic Chemistry*, 11th edn. 2014. Copyright Wiley-VCH Verlag GmbH & Co. KGaA. Reproduced with permission.)   (*Continued*)

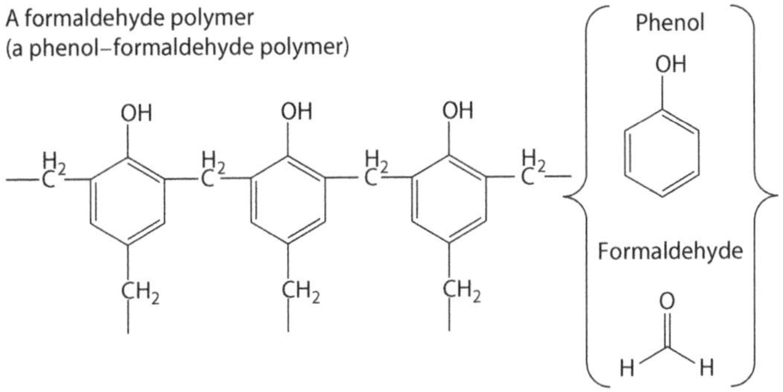

FIGURE 9.2 (Continued) Chemical structures of selected condensation polymers from Table 9.1, their corresponding monomers, and the condensation by-products.

to be developed. Because many catalysts are metal based and toxic, they too need to be replaced by the green alternatives. In Section 9.4, we provide specific examples of greening a polymerization process.

9.4 GREENING OF THE CHEMICAL INDUSTRY

In this section, we show various examples of greening of the chemical industry, including polymer industry. Examples include greening the industrial production of polycarbonates, greening the raw materials for chemical industry, and using enzymes as green catalysts. We address these topics in Sections 9.4.1 through 9.4.3.

9.4.1 GREENING THE INDUSTRIAL PRODUCTION OF POLYCARBONATES

In this section, we first address the industrial production of polycarbonates, which is based on the use of an extremely toxic reactant, phosgene (see Figure 9.2). Our goal is to impart on students that industrial processes in general, and specifically this one, involve much more than knowing a simple chemical equation, such as that from Figure 9.2. A practical implementation of a chemical reaction as described by the chemical equation requires much ingenuity and the solving of many engineering problems. After we describe the phosgene-based polycarbonate production and the associate problems (Green and Wittcoff, 2003b), we shall address the greening of the process (Green and Wittcoff, 2003c).

Phosgene, $Cl_2C{=}O$, is the acid chloride derivative of carbonic acid, $(HO)_2C{=}O$. It is much more reactive in the condensation reaction with bisphenol A (BPA) than the carbonic acid. This condensation reaction is an esterification, because ester bonds are formed. The esterification reaction is driven forward by the loss of chloride ion, Cl^-, from phosgene. The description of this reaction and its mechanism can be found in any basic organic chemistry textbook (e.g., Solomons et al., 2014). Unfortunately, phosgene is extremely toxic, so much so that it was used as a poison gas by Germany, in World War I. The toxicity of phosgene is due to its reactivity with hydroxyl or other similarly reactive functional groups from the respiratory system.

Phosgene is a gas and BPA is a solid. Thus, the contact between these two reagents is limited, and it needs to be improved. This is accomplished as follows. First, BPA, which is chemically a phenol (Ph–OH), is converted to its salt, a phenoxide (Ph–O⁻Na⁺), by dissolving it in the aqueous NaOH. Phosgene is dissolved in methylene chloride, which is a water-immiscible organic solvent. The polymerization reaction takes place at the interface of these two immiscible phases. Such a polymerization is termed interfacial. Phosgene reacts much faster with the phenoxide ion than with water, and thus is not destroyed by water in any significant amount.

The greening of the polycarbonate production requires getting rid of phosgene by designing a different process. Such a process was developed by industry, based on an old synthesis of polycarbonates, which utilized a transesterification reaction. The latter is a process in which the two esters interchange their alcohol or phenol moieties. The description of the transesterification reaction and its mechanism can be found in any basic organic chemistry textbook (e.g., Solomons et al., 2014). The elimination of phosgene makes the new process greener. Figure 9.3 shows this greened process.

In the phosgene process, the esterification reaction is driven toward the product by elimination of the chloride ion. In the transesterification reaction, the formation of polycarbonate ester is driven by the elimination of a stable phenoxide ion and by the removal of the resulting phenol by distillation under vacuum. A more detailed description of this industrial process can be found in the work of Green and Wittcoff (2003c), together with a very favorable economic analysis compared to the old phosgene-based process.

Another procedure calls for heating together BPA and diphenyl carbonate, which results in the formation of a molten polymer. The phenol is removed by distillation under reduced pressure.

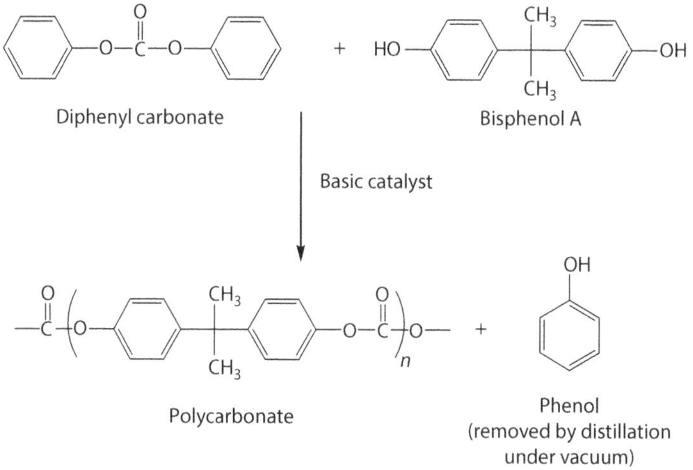

FIGURE 9.3 Synthesis of polycarbonates based on transesterification. (Green, M. M. and Wittcoff, H. A.: *Organic Chemistry Principles and Industrial Practice.* pp. 275–281. 2003c. Copyright Wiley-VCH Verlag GmbH & Co. KGaA.)

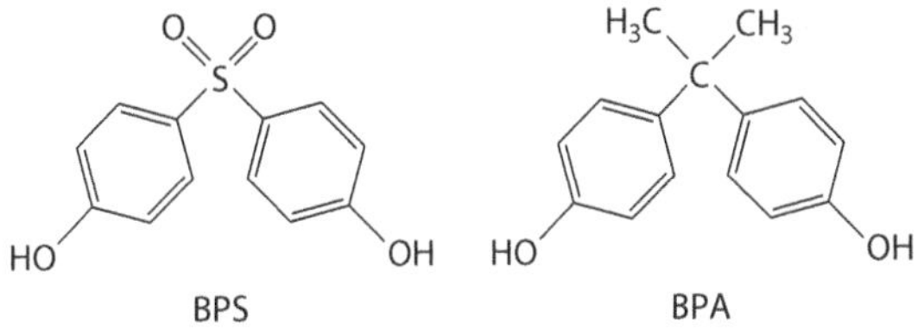

FIGURE 9.4 Structures of BPS and BPA.

In this greening process, we have focused on phosgene, which is a poisonous gas that shows acute toxicity. However, there are other toxicity types that need to be considered. Thus, BPA, especially in the literature tailored to the general public, is an endocrine disruptor that can mimic estrogen, a female hormone. Endocrine disruptors in general mimic hormonal function and could cause developmental and reproductive problems in animals and humans, and also in their offspring. This has been supported by numerous studies (Zimmerman and Anastas, 2015). Thus, BPA is not green, based on its toxicity, and also needs to be replaced. It was used in baby bottles, among other containers. Thus babies, who constitute a particularly vulnerable population for the toxic effects of BPA, could be exposed to it, because BPA may leach out of the polycarbonate container. Such leaching could happen, for example, due to any leftover BPA that is not consumed in the reaction. For this reason, the use of BPA in baby bottles was abandoned in 2012.

BPA was replaced with a sulphonyl analog, bisphenol S (BPS), which was thought to be more resistant to leaching (Bilbrey, 2014). The structure of BPS is shown in Figure 9.4, along with that of BPA, for comparison.

Unfortunately, BPS has been found to have similar toxicological concerns as BPA, namely, it is also an endocrine disruptor. Substitution of BPA with BPS is what Zimmerman and Anastas (2015) call a "regrettable substitution." As more and more becomes known about toxicological profiles of the "green" substitutes, some such substitutes may have to be replaced again.

9.4.2 GREENING RAW MATERIALS FOR CHEMICAL INDUSTRY

Chemical industry is heavily dependent on petroleum as its primary feedstock for chemicals. Because petroleum is a nonrenewable resource, greening of chemical industry needs to involve a critical step of replacing petroleum with renewable resources. It is a process that will take time, but that has already shown promise and success.

Selected examples of such greening are listed for the production of the following:

- Ethanol from plants—it has multiple uses as a solvent and a raw material for further syntheses
- Ethylene (ethene) from bioethanol (ethanol from plant sources), to make polyethylene
- Polyols (polyhydroxy compounds) from soya, to make polyurethanes
- Surfactants (detergents) from carbohydrates and oils from plants

- Biomass-derived enzymes, such as enzymes for detergents and industrial enzymes for various other purposes
- Plastics based on starch
- Polymers based on lactic acid (polylactic acid)

9.4.3 ENZYMES AS GREEN CATALYSTS

Industrial processes often involve the use of catalysts. Many such catalysts are metal based, expensive, and toxic. In many cases, they are needed not only to speed up the reaction but also to give products with desirable stereochemical and regiochemical features (spatial and positional characteristics, respectively). An example is the use of Ziegler–Natta catalysts in polymerization of alkenes. These catalysts are prepared from titanium compounds with a trialkyl aluminum as a promoter. With these catalysts, ethylene (ethene) polymerizes to give a linear polymer. When propylene (propene) polymerizes, it gives only isotactic polypropylene, rather than the other possible polymer types, such as atactic or syndiotactic. The type of polymer is related to its physical properties, and thus, one can design a desired polymer by using specific catalysts. Other metal-based catalysts can also be used.

The greening of the catalysts can be achieved by using enzymes as catalysts. Enzymes are natural catalysts that are involved in almost all biosynthetic processes in living cells. They enable chemical reactions to occur at the temperature and pressure conditions that are compatible with living organisms. They ensure high stereo- and regiospecificity of the reactions. Due to such properties, enzymes have been used to effect polymerizations, as well as other types of reactions.

Enzymes are classified by the type of reactions they catalyze in living systems. Examples include "oxidoreductases," which catalyze oxidation or reduction, "transferases," which transfer a group from one molecule to another, and "hydrolases," which perform hydrolysis in water. Enzymes from these classes are known to induce or catalyze polymerizations. For example, some oxidoreductases can be used to make polystyrene and polymethyl methacrylate, whereas some transferases and hydrolases can be used for preparation of polyesters.

Enzymes are biodegradable, which add to the green aspect of their use. There are numerous successful uses of enzymes in the pharmaceutical industry. We shall cover these in Chapter 10.

9.5 GREEN CHEMISTRY METRICS

Green chemistry metrics are an essential tool for green chemistry applications. Metrics enable us to evaluate as quantitatively as possible various factors which we have considered mostly qualitatively when we performed green analyses so far. We have already learned about atom economy as a metric. However, there are many more metrics, which encompass factors such as yields, solvents, waste, energy, process type, life cycle, cost, ease of workup, waste treatment, solvent recovery, physical and toxicological hazards, and global warming potential. We summarize these metrics in Table 9.2.

TABLE 9.2

Summary of Selected Metrics: Names, Descriptions, and Literature Sources

Name	Definitions or Descriptions	Metrics — Comments	Selected Literature Sources
Experimental atom economy	It is similar to atom economy, but it includes actual quantities of reactants/reagents that are used.	It is highly relevant when one of the reactants/reagents is used in excess.	Doxsee and Hutchinson (2004)
Percentage yield × experimental atom economy	It is obtained by multiplying the percentage yield and experimental atom economy.	It combines atom economy and percentage yield.	Doxsee and Hutchinson (2004)
Environmental factor (E-factor)	It is obtained by dividing the total mass of waste (kg) by the mass of desired product (kg).	Small values for E-factor are better than the large ones.	Doxsee and Hutchinson (2004); Van Aken et al. (2006)
Environmental quotient (EQ)	E-factor × Q, where Q is a coefficient that reflects the environmental hazard.	Q is 1 for NaCl; it is 100–1000 for heavy metals, based on their toxicity. Thus, a small Q is better than a large one.	Van Aken et al. (2006)
Effective mass yield	It is obtained by dividing the mass of desired product by the mass of all nonbenign materials used in the synthesis, expressed as the percentage.	This metric addresses toxicity and other nonbenign properties of materials.	Constable et al. (2002); Van Aken et al. (2006)
Mass intensity (MI)	It is the ratio of the total mass of materials used in a process or process step (kg) and the mass of product (kg).	This metric takes into account the yield, stoichiometry, solvents, reagents and catalysts that are used in synthesis, and chemicals that are used in workup, extractions, and crystallization. It does not include water.	Constable et al. (2002, 2009); Van Aken et al. (2006)
Process profile	It takes into account factors involved in large-scale productions, such as yield, cost, environmental and occupational hazards, raw material availability, and patent protection.	This is one of the "unified metrics," because it combines other metrics, parameters, and factors.	Van Aken et al. (2006)

(Continued)

TABLE 9.2 (*Continued*)

Summary of Selected Metrics: Names, Descriptions, and Literature Sources

Name	Definitions or Descriptions	Comments	Selected Literature Sources
Life cycle analysis (LCA)	This metric evaluates all stages of the life cycle of a product, environmental impacts of by-products and auxiliaries, such as solvents, and technical facilities.	It is often called "life cycle assessment," because it evaluates the environmental burden associated with a product, process, or activities. It is also a "unified metric."	Marteel et al. (2003); Van Aken et al. (2006)
Proprietary metrics	These are developed by the government agencies and corporations, to allow for specific analyses that fit their own needs.	Examples of parameters are the cost of waste, effluent treatment, waste disposal, nature of solvents, energy usage, and so on	Van Aken et al. (2006)
EcoScale	It is a semiquantitative tool that evaluates organic syntheses based on yield, cost, safety, reaction conditions, and ease of workup and purification. It assigns a range of penalty points to these parameters.	It can be modified by users to assign different penalty points. It is quite useful in comparing several preparations of the same product, based on safety, economical, and ecological features.	Van Aken et al. (2006)
Solvent intensity	It is obtained by dividing total solvent input (excluding water) (kg) by the total mass input (kg).	This is a useful mass indicator that addresses directly the solvent use.	Constable et al. (2009)
Waste intensity	It is obtained by dividing the total waste produced (kg) by the total mass input (kg).	This is a useful mass indicator that addresses directly the production of waste.	Constable et al. (2009)
Energy intensity	It is obtained by dividing total process energy (MJ) by the mass of final product (kg).	This energy metric addresses the overall energy expenditure of a process, relative to the mass of the final product.	Constable et al. (2009)
Life cycle energy	It is obtained by dividing the life cycle energy requirements (MJ) by the mass of final product (kg).	The life cycle energy requirement is a sum of the energies for production of raw materials, syntheses, manufacturing processes, recovery or recycling of materials, and waste treatment, as some examples.	Constable et al. (2009)

(*Continued*)

TABLE 9.2 (*Continued*)
Summary of Selected Metrics: Names, Descriptions, and Literature Sources

Name	Definitions or Descriptions	Metrics	
		Comments	Selected Literature Sources
Waste treatment energy	It is obtained by dividing energy requirement for waste treatment (MJ) by the mass of the final product (kg).	This energy metric addresses the energy needed for waste treatment relative to the mass of the final product.	Constable et al. (2009)
Solvent recovery energy	It is obtained by dividing the energy required for recovery of solvent (MJ) by the mass of the final product (kg).	This energy metric addresses the energy expenditure needed for the recovery of solvent, relative to the mass of the final product.	Constable et al. (2009)
Solvent energy ratio	It is obtained by dividing total energy for solvent use and recovery (MJ) by the total energy input (MJ).	This metric tells us how much of the total energy is devoted to the solvent use and recovery.	Constable et al. (2009)
Waste energy ratio	It is obtained by dividing the total waste produced (kg) by the total energy input (MJ).	This metric indicates the energy requirement for waste.	Constable et al. (2009).
Physical hazards metrics	They address corrosion of metals and tissues, flammability, pH extremes, and capability to cause violent reactions.	These metrics are specifically developed to address each of the physical hazards, such as rate of ignition and burning for flammability, for example.	Anastas (2008)
Toxicological hazards metrics	These address lethality, acute and chronic toxicity, and developmental and reproductive toxicity, among others. Species considered are mammalian, aquatic vertebrates and invertebrates, and plants.	These metrics are specifically developed to address the hazard in question, such as LD_{50} for lethality (a dose in grams of substance per kilograms of body weight, which results in death of 50% of animal population within 1 week of administration).	Anastas (2008)
Global warming potential metric	This metric addresses a compound's ability to absorb infrared radiation.	This metric is an example of global hazards metrics.	Anastas (2008)
Ozone depletion metric	This metric addresses a compound's ability to reach the stratosphere and interact with and destroy ozone.	In this metric, which is one of the global hazard metrics, a measurement is made to determine the atmospheric life time of a compound with the potential to deplete ozone.	Anastas (2008)

We need to familiarize ourselves with metrics and understand various definitions. This section needs to be revisited often, especially when we talk about the pharmaceutical industry in Chapter 10. A modest goal for this section is for students to understand metrics definitions sufficiently to realize that metrics actually opens the door for "quantitative greening" of chemical processes. By optimizing the existing metrics, we can improve the greenness of the chemical process, which is especially important if we do not have as yet green alternatives to the process. We provide some exercises on the use of metrics at the end of this chapter.

In Chapter 10, on pharmaceutical chemistry and its greening, we shall provide examples of applications of metrics, because such industry is notorious for multistep syntheses, which are heavy in using solvents, catalysts, and toxic chemicals in general.

Figure 9.5 lists formulas for calculations of various metrics.

Atom economy (AE) = (MW target molecule/ sum of MWs of starting materials and reagents) × 100 (%)

Experimental AE = (Theoretical yield of the product/ sum of actual weights of reactants) × 100 (%)

Environmental factor (E-factor) = Total mass of waste (kg)/mass of desired product (kg)

Mass intensity (MI) = Total mass used in a process or process step (kg)/ mass of product (kg)

Solvent intensity = Total solvent input excluding water (kg)/ total mass input (kg)

Waste intensity = Total waste produced (kg)/ total mass input (kg)

Energy intensity = Total process energy (MJ)/ mass of final product (kg)

Life cycle energy = Life cycle energy requirements (MJ)/ mass of final product (kg)

Waste treatment energy = Waste treatment energy requirement (MJ)/ mass of final product (kg)

Solvent recovery energy = Solvent recovery energy requirement (MJ)/ mass of final product (kg)

FIGURE 9.5 Formulas for calculating some commonly used metrics. (From Anastas, N. D., Incentives for using green chemistry and the presentation of an approach for green chemical design, in *Green Chemistry Metrics: Measuring and Monitoring Sustainable Processes*, Lapkin, A. and Constable, D. J. C. eds., Blackwell Publishing, Wiley, Chichester, 2008, pp. 27–40; Constable, D. J. C. et al., *Green Chem.*, 4, 521–527, 2002; Constable, D. J. C. et al., Process metrics, in *Green Chemistry Metrics: Measuring and Monitoring Sustainable Processes*, Lapkin, A. and Constable, D. J. C. eds., Blackwell Publishing, Wiley, Chichester, 2009, pp. 228–247; Doxsee, K. M. and Hutchinson, J. E., *Green Organic Chemistry: Strategies, Tools, and Laboratory Experiments*, Brooks/Cole, Toronto, Canada, 2004, pp. 89–92; Van Aken, K. et al., *Beilstein J. Org. Chem.*, 2, 1–7, 2006.)

9.6   SUMMARY

In this chapter, we gave an overview of chemical industry, with an emphasis on polymer industry, which is one of its major branches. This was necessary, because these topics are not usually covered in the basic organic chemistry textbooks, although some specific examples may be given. We have then addressed selected ways of greening of the chemical industry. Finally, we have covered metrics.

In Chapter 10, on the pharmaceutical industry and its greening, we shall expand on some topics, such as enzymatic catalysis, and will provide examples of use of metrics criteria.

REVIEW QUESTIONS

9.1 Fill in the blanks in Table 9.3.

9.2 Examine Table 9.2 and pick all the metrics that are related to the waste.

9.3 Ecological hazards, such as persistence in environment and bioaccumulation, can also be evaluated via the appropriate metrics. As a group exercise, devise the ways you could measure such hazards, based on the chemical tools that you have.

9.4 What are plastics, and how do they differ from polymers in general?

9.5 What are the differences between commodity plastics and engineering plastics?

9.6 Phosgene is very reactive with water, which destroys it. Explain the fact that phosgene can be used for the polymerization reaction with the aqueous salt of BPA without being destroyed by water.

9.7 Consider phosgene-based production of polycarbonate. Phosgene is extremely toxic, and methylene chloride, which is used in the interfacial polymerization, is toxic, volatile, and a halogenated solvent. When trying to green this process, which metrics would you consider?

9.8 A group project: Explore the references about BPA from the reference section. Start with the Wikipedia sources that are available, and add other sources that are available from your library. These sources reveal that BPA is used in epoxy resin coatings on the inside of many food and beverage cans, from which they leach and are ingested. Discuss the following question: Because BPA is toxic, how come that its use in these food and beverage containers is not prohibited by the U.S. Food and Drug Administration (FDA)?

ANSWERS TO REVIEW QUESTIONS

9.1 Consult Table 9.2 to fill in the missing contents of the second column. For the third column, use either the quantitative relationships or a qualitative assessment of what is green, if such a relationship is not described. Examples are

TABLE 9.3

Table on Metrics for Question 9.1

Name	Definitions or Descriptions	The Larger Number for the Metric Is Greener (Enter "Yes" or "No," or "Cannot Tell," or "Not Applicable," as Appropriate)
Experimental atom economy		
Environmental factor (E-factor)		
Environmental quotient (EQ)		
Mass intensity (MI)		
Proprietary metrics		
Solvent intensity		
Energy intensity		
Waste treatment energy		
Solvent recovery energy		

(Continued)

TABLE 9.3 (*Continued*)
Table on Metrics for Question 9.1

Name	Metrics Definitions or Descriptions	The Larger Number for the Metric Is Greener (Enter "Yes" or "No," or "Cannot Tell," or "Not Applicable," as Appropriate)
Waste energy ratio		
Toxicological hazards metrics		
Global warming potential metric		
Ozone depletion metric		

as follows: waste is not green, energy expenditure is not green, flammability is hazardous, the more flammable the less green, and so on.

9.2 Examine Table 9.2.

9.3 Answers will vary. Examples may include the following: collect soil and measure the amounts of pesticides in them; measure the rate of decomposition of pesticides upon heating, treatment with acid or base, or UV light; measure solubility of a pesticide in the fat, such as lard, as a model for accumulation in tissues; and so on.

9.4 Plastics are typically polymer based, but often have additives that make them more malleable, as needed for a desired application, to improve performance in general, or to reduce cost.

9.5 Typically, commodity plastics are high volume and low cost, whereas engineering plastics are low volume and high cost.

9.6 Phosgene resides most of the time in the methylene chloride phase and thus has only a limited contact with water. When such contact occurs, it is at the interface with the aqueous phase, which contains BPA salt. When exposed to water and BPA salt, phosgene reacts much faster with the phenoxide salt than with water, and is thus not destroyed by water in any significant amount.

9.7 There are many metrics to be considered, and the answers will vary. Examples are as follows: toxicological hazards metrics, environmental quotient, solvent intensity, and the process profile.

9.8 The FDA is not saying that BPA is not toxic. Instead, it states that BPA is safe at the current levels occurring in foods. Remember that toxicity is related to the dose.

REFERENCES

Anastas, N. D. (2008). Incentives for using green chemistry and the presentation of an approach for green chemical design, in *Green Chemistry Metrics: Measuring and Monitoring Sustainable Processes*, Lapkin, A. and Constable, D. J. C., eds., Blackwell Publishing, Wiley, Chichester, pp. 27–40.

Andraos, J. (2009). Applications of green metrics analysis to chemical reactions and synthesis plans, in *Green Chemistry Metrics: Measuring and Monitoring Sustainable Processes*, Lapkin, A. and Constable, D. J. C., eds., Blackwell Publishing, Wiley, Chichester, pp. 69–199.

Biermann, U., Friedt, W., Lang, S., Lühs, W., Machmüller, G., Metzger, J. O., gen. Klass, M. R., Schäfer, H. J., and Schneider, M. P. (2000). New syntheses with oils and fats as renewable raw materials for the chemical industry, *Angew. Chem. Int. Ed.,* 39, 2206–2224.

Bilbrey, J. (August 11, 2014). BPA-free plastic containers may be just as hazardous, *Sci. Am.* http://www.scientificamerican.com/article/bpa-free-plastic-containers-may-be-just-as-hazardous/.

Bisphenol A: https://en.wikipedia.org/wiki/Bisphenol_A, accessed on July 24, 2015.

Bisphenol S: https://en.wikipedia.org/wiki/Bisphenol_S, accessed on July 25, 2015.

Catalysts for polymerization reactions: http://www.essentialchemicalindustry.org/polymers-an-overview.html, accessed on July 27, 2015.

Centi, G. and Perathoner, S. (2009). From green to sustainable industrial chemistry, in *Sustainable Industrial Processes*, Cavani, F., Centi, G., Perathoner, S., and Trifiró, G., eds., Wiley-VCH, Weinheim, Germany, pp. 1–72.

Chemical industry: https://en.wikipedia.org/wiki/Chemical_industry, accessed on June 4, 2015.

Constable, D. J. C., Curzons, A. D., and Cunningham, V. I. (2002). Metrics to "green" chemistry—Which are the best? *Green Chem., 4*, 521–527.

Constable, D. J. C., Jimenez-Gonzalez, C., and Lapkin, A. (2009). Process metrics, in *Green Chemistry Metrics: Measuring and Monitoring Sustainable Processes*, Lapkin, A. and Constable, D. J. C., eds., Blackwell Publishing, Wiley, Chichester, pp. 228–247.

Dale, B. E. (2003). "Greening" the chemical industry: Research and development priorities for biobased industrial products, *J. Chem. Technol. Biotechnol., 78*, 1093–1103.

Doxsee, K. M. and Hutchinson, J. E. (2004). *Green Organic Chemistry: Strategies, Tools, and Laboratory Experiments*, Brooks/Cole, Toronto, Canada, pp. 89–92.

Eissen, M., Geisler, G., Bühler, B., Fischer, C., Hungerbühler, K., Schmid, A., and Carreira, E. M. (2009). Mass balances and life cycle assessment, in *Green Chemistry Metrics: Measuring and Monitoring Sustainable Processes*, Lapkin, A. and Constable, D. J. C., eds., Blackwell Publishing, Wiley, Chichester, 200–227.

Green, M. M. and Wittcoff, H. A. (2003a). *Organic Chemistry Principles and Industrial Practice*, Wiley-VCH, Weinheim, Germany, p. 249.

Green, M. M. and Wittcoff, H. A. (2003b). *Organic Chemistry Principles and Industrial Practice*, Wiley-VCH, Weinheim, Germany, pp. 96–101.

Green, M. M. and Wittcoff, H. A. (2003c). *Organic Chemistry Principles and Industrial Practice*, Wiley-VCH, Weinheim, Germany, pp. 275–281.

Hogson, S. C., Bigger, S. W., and Billingham, N. C. (2001). Studying polymers in the undergraduate chemistry curriculum, *J. Chem. Ed., 78*, 553–556.

Kralisch, D. (2009). Application of LCA in process development, in *Green Chemistry Metrics: Measuring and Monitoring Sustainable Processes*, Lapkin, A. and Constable, D. J. C., eds., Blackwell Publishing, Wiley, Chichester, pp. 248–271.

Lintelmann, J., Katayama, A., Kurihara, N., Shore, L., and Wenzel, A. (2003). Endocrine disruptors in the environment, *Pure Appl. Chem., 75*, 631–638.

Loos, K., ed. (2011). *Biocatalysis in Polymer Chemistry*, Wiley-VCH, Weinheim, Germany.

Marteel, A. E., Davies, J. A., Olson, W. W., and Abraham, M. A. (2003). Green chemistry and engineering: Drivers, metrices and reduction to practice, *Annu. Rev. Environ. Resour., 28*, 401–428.

Mercer, S. M., Andraos, J., and Jessop, P.G. (2012). Choosing the greenest synthesis: A multivariate metric green chemistry exercise, *J. Chem. Ed., 89*, 215–220.

Mülhaupt, R. (2013). Green polymer chemistry and bio-based plastics: Dreams and reality, *Macromol. Chem. Phys., 214*, 159–174.

Plastic: https://en.wikipedia.org/wiki/Plastic, accessed on June 4, 2015.

Polycarbonates: http://www.essentialchemicalindustry.org/polymers/polycarbonates.html, accessed on July 27, 2015.

Polymer: https://en.wikipedia.org/wiki/Polymer, accessed on June 4, 2015.

Ribeiro, M. G. T. C. and Machado, A. A. S. C. (2013). Holistic metrics for assessment of the greenness of chemical reactions in the context of chemical education, *J. Chem. Ed., 90*, 432–439.

Rosen, S. L. (1993). *Fundamental Principles of Polymeric Materials*, 2nd edn., John Wiley & Sons, New York.

Rosenmai, A. K., Dybdahl, M., Pedersen, M., van Vugt-Lussenburg, B. M. A., Wedebye, E. B., Taxvig, C., and Vinggaard, A. M. (2014). Are structural analogues to bisphenol A safe alternatives? *Toxicol. Sci., 139*, 35–47.

Rubin, B. S. (2011). Bisphenol A: An endocrine disruptor with widespread exposure and multiple effects, *J. Steroid Biochem. Mol. Biol., 127*, 27–34.

Solomons, T. W. G., Fryhle, C. B., and Snyder, S. A. (2014). *Organic Chemistry*, 11th edn., John Wiley & Sons, Hoboken, NJ.

Stevens, M. P. (1999). *Polymer Chemistry: An Introduction*, 3rd edn., Oxford University Press, New York.

Stine, W. R. (1994). *Applied Chemistry*, 3rd edn., D. C. Heath and Company, Lexington, MA.

Tao, J. and Kazlauskas, R. J., eds. (2011). *Biocatalysis for Green Chemistry and Chemical Process Development*, John Wiley & Sons, Hoboken, NJ.

Tibone, M. D., Cregg, J. J., Beckman, E. I., and Landis, A. E. (2010). Sustainability metrics: Life cycle assessment and green design in polymers, *Environ. Sci. Technol., 44*, 8264–8269.

Van Aken, K., Strekowski, L., and Petiny, L. (2006). EcoScale, a semi-quantitative tool to select and organic preparation based on economical and ecological parameters, *Beilstein J. Org. Chem.*, 2(3), 1–7.

Zimmerman, J. B. and Anastas, P. T. (2015). Towards substitution with no regrets, *Science*, 347, 1198–1199.

10 Applications of Green Chemistry Principles in the Pharmaceutical Industry

Pharmaceutical Green Chemistry is the ideal that one strives for, and the pursuit of this ideal will lead to ever better process chemistry.

...the real driver of Pharmaceutical Green Chemistry is synthetic efficiency.

John L. Tucker (2006)

It is worth noting at this juncture that "greenness" is a relative description and there are many shades of greenness.

Roger Sheldon (2010)

LEARNING OBJECTIVES

The learning objectives for this chapter are as follows.

Learning Objectives	Section Numbers
Become familiar with the basics of the pharmaceutical industry	10.1
Learn about inherent difficulties of the pharmaceutical industry in meeting green chemistry goals	10.2
	10.2.1
	10.2.2
Learn about the greening of the pharmaceutical industry, including specific examples	10.3
	10.3.1
	10.3.2
	10.3.3

10.1 INTRODUCTION TO THE PHARMACEUTICAL INDUSTRY

The pharmaceutical industry designs, develops, produces, and markets medicinal and related products used to diagnose, cure, treat, or prevent diseases.

At the foundation of drug design are medicinal and pharmaceutical chemistry. In addition to the knowledge of organic synthesis, drug designers must have a thorough understanding and expertise in various aspects of pharmaceutical sciences. The latter inform them about (1) biochemical and physiological effects of drugs; (2) the ways drugs concentrate at various sites in the body and are metabolized; (3) harmful, toxic, or otherwise undesirable side effects of drugs; and (4) the mode of action of drugs, for example, if they are binding to particular receptors in the body. Thus, drug design is typically a multidisciplinary effort, which is best realized as a team effort.

The pharmaceutical industry has its roots in naturally found drugs. Examples of drugs that are natural products are numerous (e.g., Walsch, 2003) as shown in the following text.

Many drugs come from botanical sources. Examples include morphine, an analgesic and narcotic, from opium poppy (Papaver somniferum); digitalis, a cardiac stimulant from foxgloves; quinine, an antimalarial, from cinchona bark; reserpine, an antihypertensive drug, from *Rauwolfia serpentina*; and Taxol®, an anticancer agent, from the bark of the Pacific yew tree.

Antibacterial drugs come from fungi and bacteria. Penicillins and cephalosporins come from fungi. Fungal metabolites provide the antifungal drug griseofulvin, the cholesterol-lowering drug lovastatin, and immunosuppressant cyclosporine.

Venom from the Brazilian arrowhead viper served as a basis for the development of captopril, a drug used to treat hypertension and congestive heart failure. Captopril was the first orally active angiotensin-converting enzyme inhibitor.

Many drugs were isolated from animal sources, such as insulin, steroid (sex) hormones, and adrenaline. Other drugs were isolated from microorganisms such as those found in soil and marine sources.

Structures of selected drugs from natural sources are shown in Figure 10.1. The pharmaceutical industry advanced with the development of modern synthetic organic chemistry, which enabled chemists to synthesize many natural products, especially those that were scarce in nature or difficult and expensive to isolate. Chemists systematically varied structures of naturally found drugs with the objectives of enhancing beneficial drug properties, diminishing undesirable side effects including toxicity, enabling oral bioavailability, increasing metabolic stability, and formulating drug for a prolonged action in the body. Chemists also designed new drugs by following the leads from nature and from the increasing knowledge of the mechanism by which drugs elicit their biological response.

As chemists synthesized a large number of structural analogs of known drugs and tested their biological activities, they developed an understanding of structure–activity relationship (SAR). SAR became a valuable tool for rational drug design. An example of SAR is given in Figure 10.1 for Taxol. The groups that are required for the activity are oxetane (the four-membered ring with O) or a similar small ring analog, Ph group or a close analog on carbon next to NH, and the OH group or a hydrolyzable ester next to that carbon. In contrast, the activity is only slightly reduced by removal of the OH or acetyl group from the eight-membered ring.

As computational chemistry advanced, chemists started using computational methods to model and visualize drug molecules. This new tool was especially useful for drugs that bind to the specific receptors in the body. Such binding then elicits a biological response. Examples of such drugs include morphine-type opioids and steroidal hormones, for example, the female hormone estrogen. SAR and computational modeling combined allow chemists to design drugs in a rational way, before they embark on lengthy and costly laboratory synthesis of potential new drug candidates.

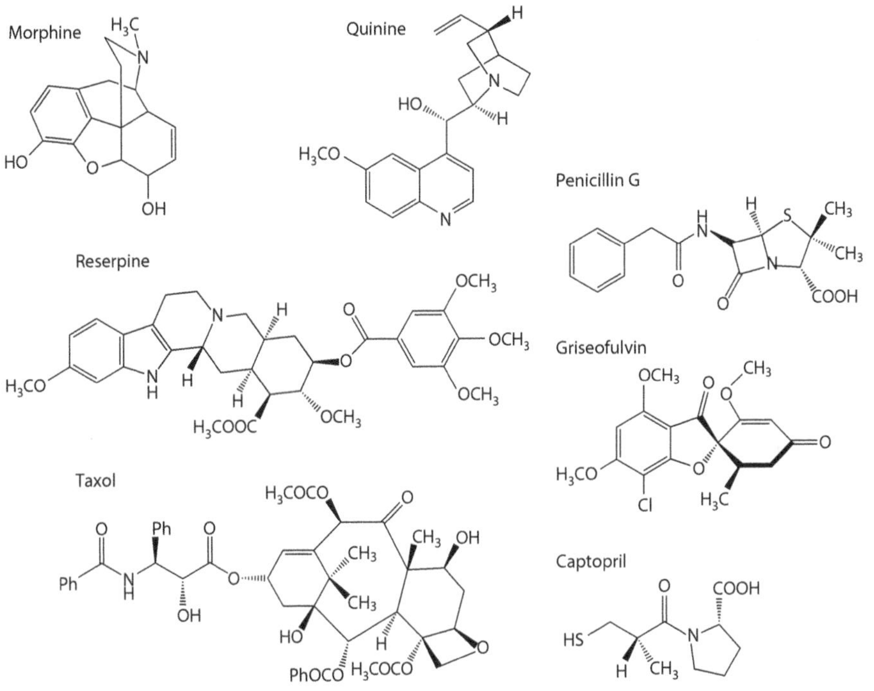

FIGURE 10.1   Structures of some common drugs from natural sources. (Thomas, G.: *Fundamentals of Medicinal Chemistry*. 2003. Copyright Wiley-VCH Verlag GmbH & Co. KGaA; *The Organic Chemistry of Drug Design and Drug Action*, 2nd edn., R. B. Silverman, Copyright 2004, with permission from Elsevier; Medicinal chemistry: https://en.wikipedia. org/wiki/Medicinal_chemistry, accessed on August 3, 2015.)

The pharmaceutical industry has some inherent difficulties in meeting green chemistry goals. However, its greening is possible, and there have been many success stories. Sections 10.2 and 10.3 address these issues and will provide specific examples.

10.2   PHARMACEUTICAL INDUSTRY OFTEN HAS DIFFICULTIES IN MEETING GREEN CHEMISTRY GOALS

Syntheses, structural modification, and derivatization of drugs are typically very elaborate processes. Syntheses often involve many steps, which may lead to overall poor atom economy. Many syntheses require protection/deprotection of functional groups (see Section 2.2.8). Specialty catalysts are often used, many of which are based on toxic metals. In addition, syntheses often have a requirement for chiral purity (thus, only one enantiomer of a possible two). In general, a thorough purification of the product is required, because impurities may have undesired biological effects. Purification typically requires use of copious amounts of solvents, many of which are toxic. All these requirements drive up the cost of drugs, cause violation of many principles of green chemistry, and result in unfavorable green chemistry metrics. Sections 10.2.1 and 10.2.2 give some examples.

10.2.1 SYNTHESIS OF PHLOROGLUCINOL, A PHARMACEUTICAL INTERMEDIATE, IS NOT GREEN ON MULTIPLE ACCOUNTS

Sheldon (2011) described the synthesis of phloroglucinol, a pharmaceutical intermediate, which is not green for multiple reasons. Chemical equations for this synthesis are shown in Figure 10.2.

This process was used for the manufacturing of phloroglucinol in the 1980s. It did meet the requirement for a good yield. Indeed, the overall yield of more than 90% was achieved over three steps.

We have learned in Section 2.2.2 about atom economy. Therefore, let us calculate atom economy of this synthesis by using the balanced equation and the formula for atom economy (Section 2.2.2 or Figure 9.5). When we do so, we find the atom economy of only 5.5%! (The details of calculations are shown as the answer to review question 10.1). The reason for the poor atom economy is a large amount of waste, comprising inorganic compounds.

In addition, this process called for an excess of reagents, including a large excess of sulfuric acid, which later has to be neutralized by base. The atom economy metric is not designed to cover such a scenario, but the environmental factor (E-factor) is (see Table 9.2 and Figure 9.5). The E-factor represents the actual amount of waste. It is obtained by dividing the total mass of waste (in kg) by the mass of desired product (in kg). Sheldon (2011) calculated the E-factor for synthesis of phloroglucinol to be 40, which means that 40 kg of waste is produced per kilogram of the desired product. The larger the E-factor is, the worse the impact on environment. The E-factor for the pharmaceutical industry in general is large, compared to other industry branches. For example, the typical E-factor for oil refining is <0.1, for bulk chemicals <1–5, for fine chemicals 5–50, and for pharmaceuticals 25–100 (Sheldon, 2000). The current aspiration target for the E-factor for pharmaceuticals is 5–50 and for fine chemicals 1–5 (Dunn et al., 2010b).

Synthesis of phloroglucinol had other green problems, such as the explosive nature of trinitrotoluene. However, the main problem for this process was the prohibitive

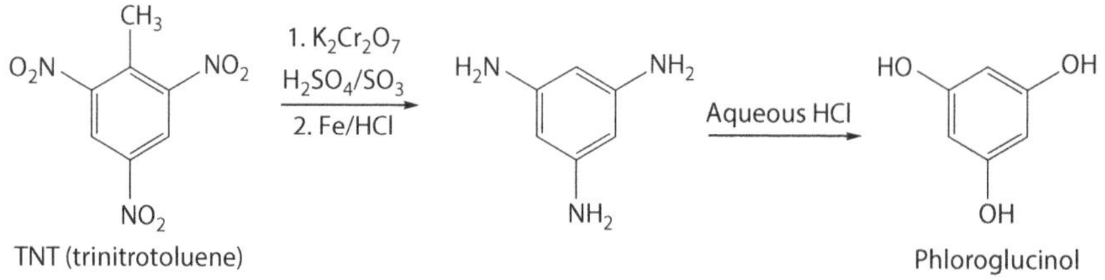

FIGURE 10.2 Chemical equations for synthesis of phloroglucinol, a pharmaceutical intermediate. (Sheldon, R. A.: Reaction efficiencies and green chemistry metrics of biotransformations. *Biocatalysis for Green Chemistry and Chemical Process Development.* 2011. 67–88. Tao, J. and Kazlauskas, R., eds. Copyright Wiley-VCH Verlag GmbH & Co. KGaA. Reproduced with permission.)

cost of the disposal of chromium-containing waste. For this reason, the process was eventually discontinued.

In Section 10.2.2, we look at another factor that potentially limits the greening of a pharmaceutical product.

10.2.2 REQUIREMENTS FOR SELECTIVITY IN SYNTHESIS OF PHARMACEUTICALS OFTEN RESULT IN UNFAVORABLE GREEN CHEMISTRY METRICS

There are different types of selectivity in organic synthesis. The first one is "chemoselectivity," which reflects competition in reactivity between functional groups. For example, one group may be preferentially reduced with selected reagent under specified conditions. The second one is "regioselectivity," which is a selective formation of one regioisomer, such as *para* versus *ortho* substituent in the aromatic ring. The third type of selectivity is "diastereoselectivity." It results in a selective formation of one diastereoisomer. The fourth type is "enantioselectivity," which provides a selective formation of one of a pair of enantiomers (Sheldon, 2010). Unless these types of selectivity are exclusive, undesired by-products will be obtained, which will create waste. A quick inspection of Figure 10.1, which shows structures of some common drugs from natural sources, reveals the presence of chiral centers, which are often multiple. The structures shown are single enantiomers. Yet, regular chemical synthesis typically yields a racemate, namely, an equal mixture of both enantiomers. In many cases, only one enantiomer has the desired biological activity, whereas the other one may be toxic or may have otherwise undesirable activity. Even if the other enantiomer is biologically inert, it would still have to be metabolized and excreted, which places a burden on the body.

The drugs from Figure 10.1 are from the natural sources, and thus are made biologically. Such compounds have proper chiral properties, because the required selectivity is achieved via enzymes, which are highly specific biological catalysts. In the laboratory, the synthesis of these and many other natural products and their analogs is not perfect. Thus, instead of the desired enantioselectivity, we would initially get a racemic mixture, as mentioned earlier. We would then have to purify such a mixture to get the biologically relevant desired enantiomer.

Meyer et al. (2009) consider six general consequences of using racemic compounds as drugs: (1) enantiomers have equal pharmacological activity; (2) one enantiomer is biologically active, whereas the other is inactive and innocuous; (3) one enantiomer is biologically active, whereas the other is toxic; (4) the two enantiomers have unequal degrees of the same activity; (5) the two enantiomers exhibit different types of pharmacological activities; and (6) the two enantiomers vary in degrees of pharmacological action and tissue specificity.

In all the cases but the first one, racemic drug formulation creates difficulties from the point of view of toxicity and safety. Thus, the trend is to try to get pure enantiomers. As an example, in 2006, 80% of small-molecule drugs (as opposed to large-molecule drugs such as proteins) that were approved by the U.S. Food and Drug Administration were chiral and 75% were single enantiomers.

There are four basic methods that are used to obtain enantiomerically pure drugs: (1) synthesis of racemate followed by chemical or physical separation of enantiomers;

(2) synthesis of racemate followed by separation of the enantiomers via a biocatalytic route; (3) asymmetric synthesis that uses catalysts with chiral ligands; and (4) asymmetric synthesis that uses enantio-, regio-, or stereoselective enzymes (or microbial whole cells that contain such enzymes) (Meyer et al., 2009). All these options are viable, depending on the type of synthesis, but the recent emphasis has been on methods 3 and 4. This will be discussed further in Section 10.3.2. All of the above steps have the potential of greatly influencing the green nature of the synthesis.

10.3　GREENING OF THE PHARMACEUTICAL INDUSTRY

The pharmaceutical industry is trying to improve on its green chemistry metrics such as E-factor, mass intensity, solvent intensity, waste intensity, and energy intensity (see Table 9.2). In addition, it is trying to comply with as many green and green engineering principles as possible. An excellent coverage of these efforts is provided in the book edited by Dunn et al. (2010a). We give several examples of greening of pharmaceutical reactions in Sections 10.3.1 through 10.3.3.

10.3.1　Greening the Synthesis of Rabeprazole, an Antiulcer Drug

Rabeprazole is an antiulcer drug, which is available worldwide under many brand names. Its traditional synthesis is presented in Figure 10.3. It comprises an oxidation of the sulfide (–S–) precursor of rabeprazole, to the sulfoxide (–S=O), which is rabeprazole, with *meta*-chloroperbenzoic acid (MCPBA) (Bhattacharya and Bandichhor, 2010).

MCPBA is a reagent that we have previously introduced in another application, namely, epoxidation of cholesterol (Section 2.2.2; Figure 2.3). In this application, it oxidizes a sulfide to a sulfoxide, but it also gives the undesired product, a sulfone (–SO_2; not shown), which results from further oxidation of sulfoxide. Due to this problem, the amount of the oxidizing reagent and the reaction conditions must be carefully controlled. However, it is difficult to avoid formation of sulfone, because the oxidation step to sulfoxide requires relatively high energy. The formation of the sulfone waste is one of the multiple reasons why the MCPBA process is not green. Thus, the yield of the MCPBA oxidation is low, only 45%. The isolation of the product is cumbersome. The MCPBA reagent itself is not green. It is expensive and shock-sensitive, and it generates lots of waste. For every gram of oxygen that is added to the sulfide to produce the corresponding sulfoxide, 10 g of *meta*-chlorobenzoic acid (MCBA) waste is produced. The atom economy is 70%. You will confirm these numbers by doing review problems 10.2 and 10.3.

The greening of this oxidation was achieved by using NaOCl oxidation. When dissolved in water, NaOCl is common household bleach. This is shown in Figure 10.4 (Bhattacharya and Bandichhor, 2010).

This reaction is green on several accounts. It gives a yield that is 76%, compared to 45% for the MCPBA oxidation. It minimizes production of the sulfone by-product. The waste product, NaCl, is environmentally acceptable. The use of NaOCl is cost effective: it costs $0.009/mole, compared to $8.65/mole for MCPBA (Bhattacharya and Bandichor, 2010). Atom economy is 86%, which is higher than 70% for the

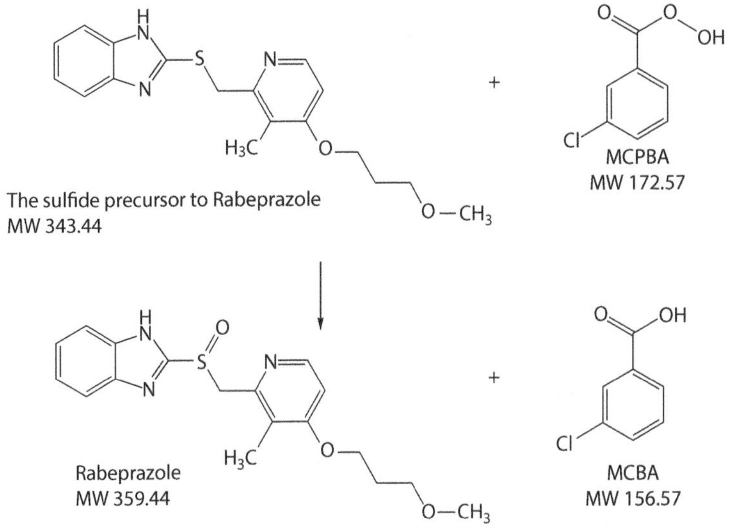

FIGURE 10.3   The traditional synthesis of rabeprazole by an oxidation of its sulfide precursor with MCPBA. (Bhattacharya, A. and Bandichhor, R.: Green technologies in the generic pharmaceutical industry. *Green Chemistry in the Pharmaceutical Industry.* 2010. 289–309. Dunn, P. J. et al., eds. Copyright Wiley-VCH Verlag GmbH & Co. KGaA.)

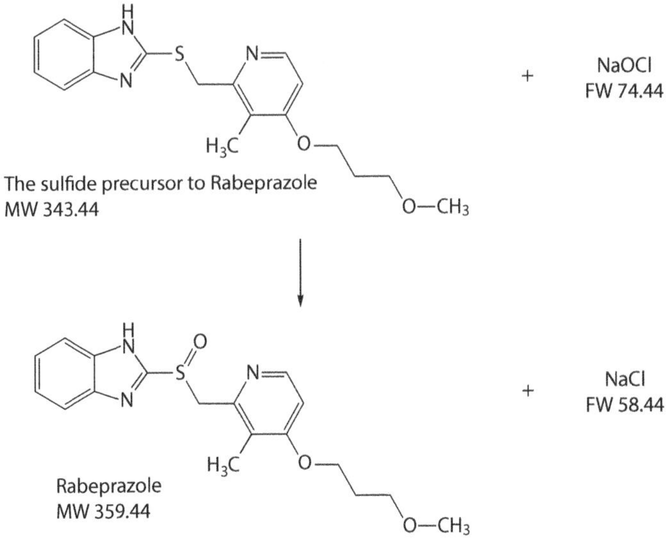

FIGURE 10.4   Green synthesis of rabeprazole by an oxidation of its sulfide precursor with NaOCl. MW, molecular weight; FW, formula weight. (Bhattacharya, A. and Bandichhor, R.: Green technologies in the generic pharmaceutical industry. *Green Chemistry in the Pharmaceutical Industry.* 2010. 289–309. Dunn, P. J. et al., eds. Copyright Wiley-VCH Verlag GmbH & Co. KGaA.)

MCPBA oxidation. In this example, the greening of the synthesis was achieved by introducing a greener oxidant.

10.3.2 TWO GREEN SYNTHESES OF L-3,4-DIHYDROXYPHENYLALANINE, A DRUG FOR TREATMENT OF PARKINSON'S DISEASE

L-3,4-Dihydroxyphenylalanine (L-DOPA) is a rare amino acid made by the human body. Its structure is shown in Figure 10.5. It is a metabolic precursor to neurotransmitters such as dopamine and adrenaline. It can be manufactured and is used as a drug for treatment of Parkinson's disease. The human body makes only L-DOPA, but not its counterpart with opposite chirality, D-DOPA.

The central problem in synthesizing L-DOPA is that enantioselectivity needs to be achieved. Only L-DOPA should be obtained, if possible. This would make the synthesis green. If both enantiomers are obtained, separation and isolation of the desired enantiomers is cumbersome and requires many steps and various materials, including solvents, which is not green.

The green synthesis can be achieved by the use of a chiral transition metal complex as a catalyst for the hydrogenation of the nonchiral precursor. The catalyst is a cationic rhodium complex, containing a chelating diphosphine ligand Ethane-1,2-diylbis[(2-methoxyphenyl)phenylphosphane] (DiPAMP) with two chiral phosphorous atoms. The resulting hydrogenation product is chiral, with a high enantioselectivity (95% enantiomeric excess). The hydrolysis removes protecting groups and gives L-DOPA. Dr. William S. Knowles, from Monsanto Company in St. Louis, Missouri, shared the 2001 Nobel Prize for the discovery and application of this catalyst (Ahlberg, 2001; Knowles, 2001). The reaction is shown in Figure 10.6.

There are many more examples of successful use of chiral catalysts in synthesis of pharmaceuticals or fine chemicals in general. Examples include catalytic asymmetric hydrogenation in the synthesis of (S)-naproxen, an anti-inflammatory drug; levofloxacin, an antibacterial agent; and a building block of vitamin E. Chirally catalyzed oxidation reactions, specifically Sharpless asymmetric epoxidation, are also used in pharmaceutical syntheses, for example, in the synthesis of β-blockers, which are used as a heart medicine. The discoveries of these chiral catalysts led to the Nobel Prize award, shared by Knowles, Noyori, and Sharpless in 2001 (Ahlberg, 2001).

L-DOPA can also be synthesized with the help of enzymes, which are natural catalysts that give the correct enantiomer exclusively. One can use either the isolated enzymes or the whole cells that contain enzymes. An industrialized bioprocess for the production of L-DOPA uses the enzyme β-tyrosinase (also called

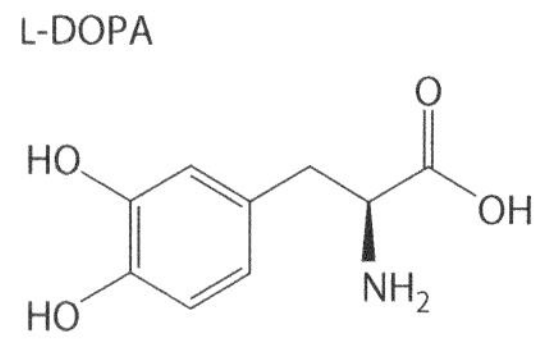

L-DOPA

FIGURE 10.5   Structure of L-DOPA.

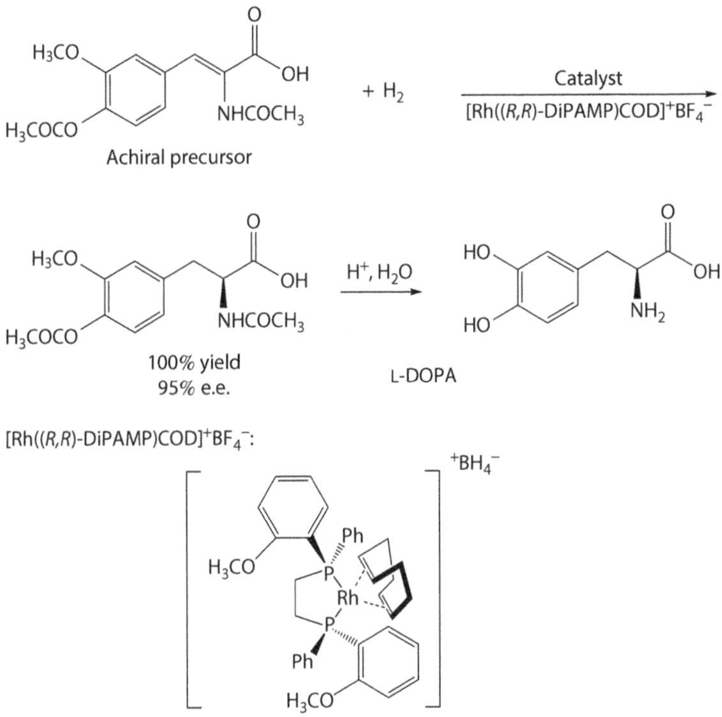

FIGURE 10.6 Green synthesis of L-DOPA by using catalytic asymmetric hydrogenation. (From Ahlberg, P., Catalytic asymmetric synthesis, Advance information on the Nobel Prize in Chemistry 2001, www.nobelprize.org/nobel_prizes/chemistry/laureates/2001/advanced-chemistryprize2001.pdf, 2001; Knowles, W. S., Asymmetric hydrogenations, Nobel Lecture, www.nobelprize.org/nobel-prizes/chemistry/laureates/2001/knowles-lecture.pdf, December 8, 2001. With permission.)

tyrosine phenol lyase), in a resting cell system of *Erwinia herbicola* (a bacterium associated with plants). This process is shown in Figure 10.7 (Meyer et al., 2009).

There are numerous other cases of using enzymatic processes to produce drugs. Examples include synthesis of Pregabalin (Lyrica), a drug for treatment of neuropathic pain; statins, such as Crestor and Lipitor, which reduce cholesterol

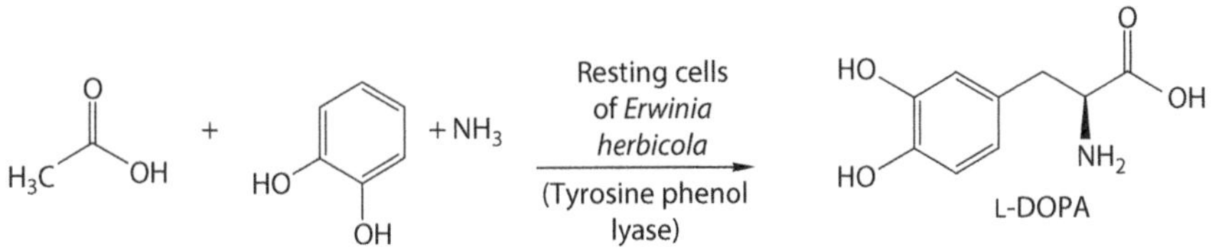

FIGURE 10.7 Enzymatic synthesis of L-DOPA. (Meyer, H.-P. et al.: Biotransformations and the pharma industry. *Handbook of Green Chemistry.* 2009. 171–212. Volume 3: Biocatalysis. Crabtree, R. H., ed. Copyright Wiley-VCH Verlag GmbH & Co. KGaA.)

levels; and 6-aminopenicillanic acid (6-APA), a key intermediate in synthesis of antibiotics (Dunn, 2010).

Green advantages of using enzymes are numerous. As one example, enzymatic hydrolysis of penicillin G to 6-APA proceeds in one step in water at 37°C, whereas a chemical process requires three steps at −40°C, various toxic reagents, and the use of toxic dichloromethane as solvent (Sheldon, 2000; Dunn, 2010).

10.3.3 Some Other Examples of Greening of Pharmaceutical Processes

Green synthesis of sildenafil citrate, marketed as Viagra, a drug for treatment of male erectile dysfunction, was so successful that it received 2003 UK Award for Green Chemical Technology (Dunn et al., 2004). The greening consisted of discovering an efficient seven-step synthesis, which did not require extraction in the workup in any steps, and implementing an efficient solvent recovery. This resulted in an impressive E-factor of 6, namely, just 6 kg of waste per kilogram of product, compared to the industry average of 25 to more than 100.

Synthesis of ibuprofen, a nonsteroidal anti-inflammatory drug, marketed as, for example, Advil®, was successfully greened. Instead of six stoichiometric steps and less than 40% atom utilization in the old process, the new innovative process features three catalytic steps and ~80% atom utilization. The latter becomes almost 99% when the recovered acetic acid is included. In the new process, anhydrous hydrogen fluoride is used as both a catalyst and a solvent. It is recovered and recycled with greater than 99.9% efficiency. There is virtually no waste generated in this process. For these reasons, this new process was recognized by the Presidential Green Chemistry Challenge Award in 1997 (see the Environmental Protection Agency [EPA] website).

Greening of the pharmaceutical processes occurs in general if chemical synthesis is designed such that it does not require isolation and purification of intermediates, but, instead, occurs in a single reactor (one pot). Such a process is named "telescoped." Isolation of intermediates is costly, and it also leads to the loss of valuable material. On a manufacturing scale, isolation of intermediates along with the isolation of the final product requires ~50% increase in labor cost and ~70% increase in equipment expenditure, as compared to the one-pot process. A further problem with isolation of intermediates is that the workers may be exposed to the pharmacologically active compounds (Zhang and Cue Jr., 2012).

Many drug intermediates have been synthesized by telescoped processes. For example, synthesis of a diazepine intermediate for the drug for treatment of schizophrenia was accomplished in one-pot, three-step process.

Reactions that are especially well suited for telescoped processes are multicomponent reactions (MCRs). In these reactions, three or more starting materials react to form a product, which incorporates essentially all of the atoms of the starting materials, and thus, the MCRs have an excellent atom economy. We have already learned about one such reaction, namely, the Passerini reaction (Section 4.2.2). The Passerini reaction can be performed "on water," thus by using a green solvent. Some other MCRs can also be performed under solventless conditions. Huang et al. (2012) cite several examples of MCRs that are used in the syntheses of key precursors of important drugs, such as praziquantel for treatment of schistosomiasis ("snail fever"), olanzapine for treatment of schizophrenia and bipolar disorder, and lidocaine, a commonly used local anesthetic.

In Chapter 11, we shall address some green analytical methods, many of which are used in pharmaceutical industry. In Chapter 12, we shall discuss the impact that pharmaceuticals have on environment, after they are eliminated from the body.

REVIEW QUESTIONS

10.1 The balanced equation for the synthesis of phloroglucinol shown in Figure 10.2 is given below. Calculate the atom economy (AE) for this synthesis (Sheldon, 2011).
$$C_7H_5N_3O_6 + K_2Cr_2O_7 + 5H_2SO_4 + 9Fe + 21HCl = C_6H_6O_3 + Cr_2(SO_4)_3 + 2KHSO_4 + 9FeCl_2 + 3NH_4Cl + CO_2 + 8H_2O$$

10.2 Calculate the atom economy for MCPBA oxidation of the sulfide precursor of rabeprazole, shown in Figure 10.3.

10.3 How many grams of waste are produced in the MCPBA process for synthesis of rabeprazole for each gram of oxygen that is added to the sulfide? Use equation and molecular weights (MWs) given in Figure 10.3.

10.4 Calculate the atom economy for the NaOCl oxidation of the sulfide precursor of rabaprazole. Use the equation and MW/FW data from Figure 10.4.

ANSWERS TO REVIEW QUESTIONS

10.1 We need to first calculate molecular weights, take into account stoichiometric coefficients, and then apply the formula for atom economy (see Section 2.2.2 or Figure 9.5).
$$AE = [126/(227 + 294 + 490 + 504 + 767)] \times 100 = (126/2282) \times 100 = {\sim}5.5\%$$

10.2 $AE = [359.44/(343.44 + 172.57)] \times 100 = (359.44/516.01) \times 100 = 69.7\%$ or ~70%

10.3 MCPBA reagent: MW 172.57
MCBA waste: MW 156.57
The difference in the MWs is oxygen: $172.57 - 156.57 = 16.00$
16.00 g of oxygen/156.57 g of waste = 1.00 g of oxygen/x g of waste; $x = 156.57/16.00 = 9.78$ g or ~10 g.

10.4 $AE = [359.44/(343.44 + 74.44)] \times 100 = (359.44/417.88) \times 100 = 86\%$

REFERENCES

Ahlberg, P. (2001). Catalytic asymmetric synthesis, Advance information on the Nobel Prize in Chemistry 2001, http://www.nobelprize.org/nobel_prizes/chemistry/laureates/2001/advanced-chemistryprize2001.pdf, accessed on February 24, 2016.

Bhattacharya, A. and Bandichhor, R. (2010). Green technologies in the generic pharmaceutical industry, in *Green Chemistry in the Pharmaceutical Industry*, Dunn, P. J., Wells, A. S., and Williams, M. T., eds., Wiley-VCH, Weinheim, Germany, pp. 289–309.

Cannon, J. G. (2007). *Pharmacology for Chemists*, 2nd edn., Oxford University Press, New York.

Dunn, P. (2010). Water as a green solvent for pharmaceutical applications, in *Handbook of Green Chemistry*, Volume 5: Reactions in Water, Li, C.-J., ed., Wiley-VCH, Weinheim, Germany, pp. 363–383.

Dunn, P. J., Galvin, S., and Hettenbach, K. (2004). The development of an environmentally benign synthesis of sildenafil citrate (Viagra™) and its assessment by green chemistry metrics, *Green Chem.,* 6, 43–48.

Dunn, P. J., Wells, A. S., and Williams, M. T., eds. (2010a). *Green Chemistry in the Pharmaceutical Industry,* Wiley-VCH, Weinheim, Germany.

Dunn, P. J., Wells, A. S., and Williams, M. T. (2010b). Future trends for green chemistry in the pharmaceutical industry, in *Green Chemistry in the Pharmaceutical Industry,* Dunn, P. J., Wells, A. S., and Williams, M. T., eds., Wiley-VCH, Weinheim, Germany, pp. 333–355.

For drugs, natural products, and pharmaceutical and medicinal chemistry, L-Dopa: https://en.wikipedia.org/wiki/L-Dopa, accessed on August 21, 2015.

For Presidential Green Chemistry Challenge Awards, http://www2.epa.gov/greenchemistry/presidential-green-chemistry-challenge-winners, accessed on February 24, 2016.

Huang, Y., Yazback, A., and Dömling, A. (2012). Multicomponent reactions, in *Green Techniques for Organic Synthesis and Medicinal Chemistry,* Zhang, J. and Cue Jr., B. W., eds., John Wiley & Sons, Chichester, pp. 497–522.

Knowles, W. S. (2001). Asymmetric hydrogenations, Nobel Lecture, December 8, 2001, http://www.nobelprize.org/nobel-prizes/chemistry/laureates/2001/knowles-lecture.pdf.

Medicinal chemistry: https://en.wikipedia.org/wiki/Medicinal_chemistry, accessed on August 3, 2015.

The Merck Index. (2001). *An Encyclopedia of Chemicals, Drugs, and Biologicals,* 13th edn., Merck & Co., Whitehouse Station, NJ.

Meyer, H.-P., Chisalba, O., and Leresche, J. E. (2009). Biotransformations and the pharma industry, in *Handbook of Green Chemistry,* Volume 3: Biocatalysis, Crabtree, R. H., ed., Wiley-VCH, Weinheim, Germany, pp. 171–212.

Moores, A. (2009). Atom economy—Principles and some examples, in *Handbook of Green Chemistry,* Volume 1: Homogeneous Catalysis, Crabtree, R. H., ed., Wiley-VCH, Weinheim, Germany, pp. 1–15.

Natural product: https://en.wikipedia.org/wiki/Natural_product, accessed on August 3, 2015.

Pharmaceutical industry: https://en.wikipedia.org/wiki/Pharmaceutical_industry, accessed on August 3, 2015.

Pohl, M., Bocke, D., and Müller, M. (2009). Thiamine-based enzymes for biotransformations, in *Handbook of Green Chemistry,* Volume 3: Biocatalysis, Crabtree, R. H., ed., Wiley-VCH, Weinheim, Germany, pp. 75–114.

Rabeprazole: https://en.wikipedia.org/wiki/Rabeprazole, accessed on August 17, 2015.

Sheldon, R. (2010). Introduction to green chemistry, organic synthesis and pharmaceuticals, in *Green Chemistry in the Pharmaceutical Industry,* Dunn, P. J., Wells, A. S., and Williams, M. T., eds., Wiley-VCH, Weinheim, Germany, pp. 1–20.

Sheldon, R. A. (2000). Atom efficiency and catalysis in organic synthesis, *Pure Appl. Chem.,* 72, 1233–1246.

Sheldon, R. A. (2011). Reaction efficiencies and green chemistry metrics of biotransformations, in *Biocatalysis for Green Chemistry and Chemical Process Development,* Tao, J. and Kazlauskas, R., eds., John Wiley & Sons, Hoboken, NJ, pp. 67–88.

Silverman, R. B. (2004). *The Organic Chemistry of Drug Design and Drug Action,* 2nd edn., Elsevier, Amsterdam, the Netherlands.

Sommer, W. and Weibel, D. (2008). DIPAMP, *Aldrich ChemFiles,* 8.2, 81.

Thomas, G. (2003). *Fundamentals of Medicinal Chemistry,* John Wiley & Sons, Chichester.

Tucker, J. (2006). Green Chemistry, a pharmaceutical perspective, *Org. Res. Dev.,* 10, 315–319.

Walsh, G. (2003). *Biopharmaceuticals, Biochemistry and Biotechnology,* 2nd edn., John Wiley & Sons, Hoboken, NJ.

Zhang, J. and Cue Jr., B. W. (2012). Green process chemistry in the pharmaceutical industry: Recent case studies, in *Green Techniques for Organic Synthesis and Medicinal Chemistry,* Zhang, J. and Cue Jr., B. W., eds., John Wiley & Sons, Chichester, pp. 631–658.

11 Applications of Green Chemistry Principles in Analytical Chemistry

Green analytical chemistry is not about monitoring of environmental pollutants but rather "greening of methodologies."

Charlotta Turner (2013)

LEARNING OBJECTIVES

The learning objectives for this chapter are as follows.

Learning Objectives	Section Numbers
Application of the 12 principles of green chemistry to analytical chemistry	11.1
Green analytical chemistry and its specialized green principles	11.2
Greening of analytical chemistry	11.3
Specific examples of greening	11.3.1
	11.3.2
	11.3.3

11.1 APPLICATION OF THE 12 PRINCIPLES OF GREEN CHEMISTRY TO ANALYTICAL CHEMISTRY

The 12 principles of green chemistry (Chapter 2) were developed in response to the use and generation of hazardous substances. This was the case primarily in the field of organic synthesis, especially in conjunction with the chemical and pharmaceutical industries. Principle 11, which addresses the need for "real-time analysis for pollution prevention," is specifically concerned with analytical chemistry. However, other principles are also applicable to analytical chemistry (Keith et al., 2007), notably principle 1 (Prevention of waste), principle 5 (Eliminate solvents and auxiliaries or make them safer), principle 6 (Design for energy efficiency), principle 8 (Eliminate derivatives or make them safer), and principle 12 (Use inherently safer chemistry for accident prevention).

Keith et al. (2007) pointed out that many analytical procedures require hazardous chemicals as part of sample preservation and preparation along with instrument

135

calibration and cleaning. Often, analytical procedures create waste in larger quantities and great toxicity than that of the sample that is analyzed.

Analytical schemes may include many steps. These fall into two broad categories: The first one includes the pretreatment steps, such as digestion, extraction, drying and concentration; and the second category includes the signal acquisition techniques, such as spectroscopy, electrochemistry, or bioanalytical chemistry. There are specific ways in which these categories can be greened. They are covered in the references at the end of the chapter, for the students and other readers with an extensive background in analytical chemistry.

11.2   GREEN ANALYTICAL CHEMISTRY AND ITS SPECIALIZED GREEN PRINCIPLES

Several investigators felt that analytical chemistry is sufficiently different from other branches of chemistry that it would benefit from tailoring and supplementing the green chemistry principles to match analytical chemistry applications (Gałuszka et al., 2013). Thus, they proposed the 12 principles of green analytical chemistry (GAC). In addition to the traditional format (a list of principles), they presented them as the mnemonic SIGNIFICANCE:

S = Select a direct analytical technique
I = Integrate analytical processes and operations
G = Generate as little waste as possible and treat it properly
N = Never waste energy
I = Implement automation and miniaturization of methods
F = Favor reagents obtained from renewable source
I = Increase safety for operators
C = Carry out *in situ* measurements
A = Avoid derivatization
N = Note that the sample number and size should be minimal
C = Choose multianalyte or multiparameter method
E = Eliminate or replace toxic reagents

Analytical procedures need to implement these principles typically in six areas:

1. *Waste*: Reduce volume, treat properly.
2. *Method*: Ideally, it should be automated, with no sample treatment of derivatization (cf. green chemistry principle 8), and with all the steps fully integrated.
3. *Instrument*: It should be energy efficient and ideally miniaturized.
4. *Reagent*: It should be nontoxic, safe, and from renewable resources.
5. *Sample*: Ideally, it should be minimal in number and smallest in size.
6. *Operator*: The person should be safe.

The greening of analytical chemistry needs to encompass all these areas as much as possible.

11.3   GREENING OF ANALYTICAL CHEMISTRY

The general direction of the greening of analytical chemistry is set by the principles of green chemistry and GAC. In this section, we provide specific examples of such a greening, which are accessible to the students of organic chemistry with some exposure to analytical chemistry.

11.3.1   GREENING OF PREPARATIVE CHROMATOGRAPHY

Chromatography is a common and powerful technique for the separation of organic compounds. It is one of the basic techniques that are taught in the beginning organic laboratory. This technique is used in virtually every laboratory, for applications spanning the routine to research. Pharmaceutical industry uses chromatography for preparative separations, at all stages, including drug discovery, process development, and manufacturing (Mihlbachler and Dapremont, 2012).

Chromatography generates waste, such as unwanted fractions and used packing material. It uses large amounts of solvents. Also, it consumes energy, for example, to evaporate and recycle the solvents, or to achieve high pressure for high-pressure liquid chromatography. The most negative impact of chromatography on the environment occurs especially in the area of solvent use (Mihlbachler and Dapremont, 2012).

There are various ways of greening chromatography (Ferguson et al., 2012). The amount of solvent and its hazards are the subject of greening, specifically via three R's of GAC, namely, "reduce," "replace," and "recycle." The use of solvent can be reduced by diminishing column dimensions, such as its length and diameter. To replace solvents, GAC depends on the solvent selection guides for green chemistry. In one such guide, by the pharmaceutical corporation Pfizer Inc., New York, common solvents are grouped in the categories of undesirable, usable, and preferred. Examples of undesirable solvents are diethyl ether, dichloromethane, chloroform, and dioxane. Useable solvents are acetonitrile, cyclohexane, and tetrahydrofuran, among others. Preferred solvents are water, ethanol, propanol, acetone, and so on. This classification considers toxicity and other hazards, and the environmental impact of solvents, among other factors. (Note: You will revisit this classification by doing two review problems.)

Another trend in replacing chromatography solvents is to use supercritical fluids, which is done in supercritical fluid chromatography (SFC). SFC typically uses CO_2 as a mobile phase. The critical temperature and pressure of CO_2 are 31.1°C and 72.9 atm, respectively. By adjusting the temperature and/or pressure, the elution properties of CO_2 can be "tuned" to give the desired values. However, it is often necessary to increase the polarity of the liquid CO_2 phase, which is similar to that of hexane, if a polar compound needs to be eluted from the chromatography column. This is accomplished by adding a small amount of a polar organic solvent, such as methanol, to the supercritical CO_2. The solvent that is added is referred to as a modifier (Ferguson et al., 2012).

Pressurized hot (subcritical) water can also be used as a mobile phase in SFC. Subcritical water has tunable polarity, which at some point resembles that of acetone (see Section 5.2.1). In some cases, pressurized CO_2 can be added as a modifier to reduce the polarity of water (Keith et al., 2007; Turner, 2013).

11.3.2 Use of Plant Extracts for Metal Ion Determination

We present here an innovative approach to the sustainability aspect of GAC. In a series of papers, Harwell and coauthors (Settheeworrarit et al., 2005; Pinyou et al., 2010; Grudpan et al., 2011; Hartwell, 2012; Insain et al., 2013) showed how plant extracts can be used as natural and sustainable reagents for the determination of iron and aluminum ions, when coupled with a flow injection (FI) analysis system. The latter system enables the use of natural reagents that are not necessarily pure, because the standards go through the analysis process under the same conditions as the samples to be determined. Because extracts of the natural reagents are minimally processed and flow-based system has low volume requirements, these factors make the analysis green.

In one example, guava leaf extract was used for iron ion determination (Settheeworrarit et al., 2005). Guava (*Psidium guajava* L.) belongs to the Myrtaceae family. In Thailand, there has been some local use of guava leaves by villagers for evaluation of iron in groundwater. If the color changed to a darker shade, villagers would treat the water with alum (a naturally found aluminum sulfate salt, as a hydrate, and containing K^+ or other cations) to precipitate iron. In the modern chemical analysis, the guava leaf extract was prepared by grinding the leaves with water along with acetate buffer at pH 4.8, and then filtering the suspension. Such an extract is stable for at least 4 h, which is sufficient for the FI experiment. This iron ion determination could tolerate other ions, such as Ca^{2+}, Mg^{2+}, Co^{2+}, Ni^{2+}, and Cr^{3+}, at least up to 1:1 concentration ratio (10 ppm iron ions: 10 ppm interference ion). This tolerance allows for a wider applicability of this method.

This natural reagent allows for the substitution of various toxic reagents that are traditionally used for the determination of iron. They include phenanthroline, triazines, azo-dyes, mercaptoquinoline, and thiocyanate.

In another study by the same group (Pinyou et al., 2010), the determination of iron ion was accomplished by using extracts from green tea (*Camellia sinensis*), by a similar preparation as in the previous example, and again by using the FI analytical technique. Green tea contains various polyphenolic compounds that interact with iron ions. The analysis by using the green tea extract was successful. It was also applicable to the analysis of iron in pharmaceutical samples. It gave results that agreed well with those obtained by standard methods. Again, tolerance for other cations, such as those that are commonly present in pharmaceutical samples, was adequate.

Insain et al. (2013) found that the crude extract from Indian almond (*Terminalia catappa* L.) leaves are suitable for determination of aluminum ion, Al^{3+}, again in conjunction with FI method of analysis. A practical use of this method was the determination of aluminum ion in the wastewater from ceramic factories. The Indian almond tree is abundant in tropical areas, but also in some parts of the United States. It is rich in polyphenolic compounds, including tannins. Such polyphenols interact with metal ions, including iron. The latter does interfere with determination of aluminum ion, but not when its concentrations are low, such as in the wastewater from ceramic factories. Similarly, Cr^{3+} interferes, but less so than the iron ions, and again, its concentration in the waste water is low. Ions such as Ca^{2+}, Mg^{2+}, and Mn^{2+} do not interfere significantly.

11.3.3 GREEN CHOICES OF INSTRUMENTS BY ENERGY CONSUMPTION

Chemistry students are exposed to various analytical techniques and corresponding instruments in teaching laboratories. Teaching of specific techniques is mostly pedagogically oriented, namely, to expose the students to the technique, and to teach them how to use the instrument and how to interpret the data. In the chemical and pharmaceutical industries and their associated laboratories at all levels, from discovery to production, many more instruments are used depending on the need for specific analytical data. However, at both the student and industry levels, more than one instrument can provide the requisite information. Then a decision needs to be made which instrument should be used from the GAC point of view. We address the instrument ranking based on one GAC criterion, that of energy consumption. This ranking has been proposed by Raynie and Driver (2009) and is discussed also in several key references (e.g., Koel and Kaljurand, 2010; Gałuszka et al., 2012; Koel, 2012). Table 11.1 shows energy consumption ranking for selected instruments that organic chemistry students are typically familiar with, together with the numerical and color ranking for greenness. The latter is based on the system in which 1 or green is the best (the greenest), 2 or yellow follows, and 3 or red is not green. We learn from this table that the FTIR instrument is greener than the NMR, based on its energy consumption. Thus, it should be used preferentially, if everything else is equal. We shall show an example in Chapter 13.

TABLE 11.1

Instrument Ranking by Energy Consumption, with Numerical and Color Ranking for Greenness

Energy Consumption [kWh Per Sample]	Green Rating, Numerical (1 Is the Best) and by Color (Green Is the Best)	Instrument
0.1 or less	1; green	FTIR spectrometer, UV–Vis spectrometer, fluorescence spectrometer
1.5 or less	2; yellow	Microwave, GC, LC
More than 1.5	3; red	NMR spectrometer, GC-MS, LC-MS

Sources: Raynie, D. and Driver, J. L., Green assessment of chemical methods, in *13th Green Chemistry and Engineering Conference*, Washington, DC, 2009; Koel, M. and Kaljurand, M., *Green Analytical Chemistry*, p. 135, RSC Publication, Cambridge, 2010; Gałuszka, A. et al., Analytical eco-scale for assessing the greenness of analytical procedures, *Trends Anal. Chem.*, 37, 61–72, 2012. With permission.

Notes: FTIR, Fourier transform infrared; UV–Vis, ultraviolet–visible; GC, gas chromatograph; LC, liquid chromatograph; NMR, nuclear magnetic resonance; GC–MS, gas chromatograph–mass spectrometer; LC–MS, liquid chromatograph–mass spectrometer.

11.4   SUMMARY

In this chapter, we have provided an overview of the application of green chemistry principles to analytical chemistry. We have also shown that more specific principles are useful for GAC. Finally, we have shown some examples of the greening analytical chemistry.

REVIEW QUESTIONS

11.1 (a) In your laboratory experience with instruments, when is derivatization of samples common and what is its purpose? (b) Are the agents that are used for such derivatization green? (c) Which principle of green chemistry addresses derivatization?

11.2 Which principle of GAC is applied in the use of natural plant sources as reagents?

11.3 From Table 11.1, we see that the nuclear magnetic resonance (NMR) spectrometer uses more energy than the Fourier transform infrared (FTIR) spectrometer. Based on your actual experience with the NMR, or a textbook description of its operation, answer the following: (a) Which instrument has a higher maintenance cost and why? (b) Which instrument can handle solid samples easier and cheaper? (c) Which instrument has a restricted access and why?

11.4 Suggest reasons why diethyl ether is an undesirable solvent for chromatography of the type you used in the beginning organic laboratory (gravity column)?

11.5 Class exercise: Compile information from Wikipedia and other sources about chromatography solvents listed in Section 11.3.1. Focus on physical properties, such as boiling point, flammability, and polarity; on chemical properties, such as the ability to form peroxides upon exposure to air; on toxicity; and on the environmental impact. Based on this research, support the classification of these solvents as undesirable, useable, and preferred.

ANSWERS TO REVIEW QUESTIONS

11.1 (a) Gas chromatography (GC). Derivatization for GC is usually done to increase volatility of samples. This is done by converting polar groups, such as N–H and O–H, which decrease volatility of the sample by hydrogen bonding, to nonpolar derivatives. The latter often contain bulky nonpolar silyl groups. (b) Most of the derivatization agents are toxic or otherwise hazardous, and thus are not green. Inspect the list of derivatization agents for GC, such as the one given on the website below:

http://www.sigmaaldrich.com/analytical-chromatography/analytical-products.html?TablePage=8658816.

(c) Principle 8

11.2 F = Favor reagents obtained from renewable sources.

11.3 (a) NMR. Most NMR spectrometers are Fourier transform NMR (FTNMR) spectrometers that use superconducting magnets. The latter are cooled by liquid helium (at 4.2 K). Liquid nitrogen (at 77.4 K) is used in the outer jacket of the cooling system. These liquid gases are expensive. The FTIR requires no such expense. (b) FTIR, especially if an Attenuated total reflection attachment is available. FTNMR requires a number of special techniques/equipment, including "magic-angle" spinning, cross-polarization, and enhanced probe electronics. (c) FTNMR. People with pacemaker should not enter the room where FTNMR is housed, because strong magnetic field generated around the instrument can cause pacemakers to fail. Also note that credit cards will be erased by the strong magnetic field next to the instrument, and thus should be left outside the room. Warnings on both accounts are usually posted at the entrance door of the NMR room.

11.4 It evaporates too quickly, and its vapors may travel upward in the column. The evaporation may create pockets of air in the column. Ether is also hazardous. It is flammable and it forms peroxides upon standing, which are explosive. Peroxides may concentrate in the column as diethyl ether is repeatedly passed through it.

11.5 Individual responses may vary. Try to consolidate and double-check all the data. Produce a group report and have your instructor evaluate it.

REFERENCES

Ferguson, P., Harding, M., and Young, J. (2012). Green analytical chemistry, in *Green Techniques for Organic Synthesis and Medicinal Chemistry*, Zhang, W. and Cue Jr., B. W., eds., John Wiley & Sons, Chichester, pp. 659–683.

Gałuszka, A., Konieczka, P., Migaszewski, Z. M., and Namieśnik, J. (2012). Analytical eco-scale for assessing the greenness of analytical procedures, *Trends Anal. Chem.*, 37, 61–72.

Gałuszka, A., Migaszewski, Z. M., and Namieśnik, J. (2013). The 12 principles of green analytical chemistry and the SIGNIFICANCE mnemonic of green analytical practices, *Trends Anal. Chem.*, 50, 78–84.

Grudpan, K., Hartwell, S. K., Wongwilai, W., Grudpan, S., and Lapanantnoppakhun, S. (2011). Exploiting green analytical procedures for acidity and iron assays employing flow analysis with simple natural reagent extracts, *Talanta*, 84, 1396–1400.

Hartwell, S. K. (2012). Exploiting the potential for using inexpensive natural reagents extracted from plants to teach chemical analysis, *Chem. Educ. Res. Pract.*, 13, 135–146.

Insain, P., Khonyoung, S., Sooksamiti, P., Lapanantoppakhun, S., Jakmunee, J., Grudpan, K., Zajicek, K., and Harwell, S. K. (2013). Green analytical methodology using Indian almond (*Terminalia catappa* L.) leaf extract for determination of aluminum ion in waste water from ceramic factories, *Anal. Sci.*, 29, 655–659.

Keith, L. H., Gron, L. U., and Young, J. L. (2007). Green analytical methodologies, *Chem. Rev.*, 107, 2695–2708.

Koel, M. (2012). Energy savings in analytical chemistry, in *Handbook of Green Analytical Chemistry*, de la Guardia, M. and Garrigues, S., eds., 1st edn., John Wiley & Sons, Chichester, pp. 291–319.

Koel, M. and Kaljurand, M. (2010). *Green Analytical Chemistry*, RSC Publication, Cambridge, p. 135.

Mihlbachler, K. and Dapremont, O. (2012). Preparative chromatography, in *Green Techniques for Organic Synthesis and Medicinal Chemistry*, Zhang, W. and Cue Jr., B. W., eds., John Wiley & Sons, Chichester, pp. 590–611.

Pinyou, P., Hartwell, S. K., Jakmunee, J., Lapanantnoppakhun, S., and Grudpan, K. (2010). Flow injection determination of iron ions with green tea extracts as a natural chromogenic reagent, *Anal. Sci.*, 26, 619–623.

Raynie, D. and Driver, J. L. (2009). Green assessment of chemical methods, in *13th Green Chemistry and Engineering Conference*, Washington, DC. acs.confex.com, paper69446_5. pdf; http://acs.confex.com/acs/green09/recordingredirect.cgi/id/457, accessed on February 24, 2016.

Settheeworrarit, T., Hartwell, S., K., Lapananoppakhun, S., Jakmunee, J., Christian, G. D., and Grudpan, K. (2005). Exploiting guava leaf extract as an alternative natural reagent for flow injection determination of iron, *Talanta*, 68, 262–267.

Tobiszewski, M., Marć, M., Gałuszka, A., and Namieśnik., J. (2015). Green chemistry metrics with special reference to green analytical chemistry, *Molecules*, 20, 10928–10946.

Turner, C. (2013). Sustainable analytical chemistry—More than just being green, *Pure Appl. Chem.*, 85, 2217–2229.

12 Application of Green Chemistry Principles in Environmental Chemistry

Some background in environmental chemistry should be part of the training of every chemistry student. The ecologically illiterate chemist can be a very dangerous species.

More often than not, it is impossible to come up with a simple answer to an environmental chemistry problem. But, building on an ever-increasing body of knowledge, the environmental chemist can make educated guesses as to how environmental systems will behave.

Stanley E. Manahan (1991)

Dealing effectively and responsibly with complex problems within complex environmental systems requires...a continuous process of critical thinking, problem solving and decision making.

Uri Zoller (2005)

Environmental science and green chemistry can collectively inform innovation.

Richard T. Williams and Travis R. Williams (2012a)

LEARNING OBJECTIVES

The learning objectives for this chapter are as follows.

Learning Objectives	Section Numbers
Learn about the objectives of environmental chemistry	12.1
Learn about the relationship between the environmental and green chemistry, and how environmental chemistry informs and challenges green chemistry	12.2
Learn about a complex environmental chemistry problem: pharmaceuticals as emerging environmental pollutants	12.3
Learn about development of ecofriendly chemicals in response to the findings of environmental chemistry	12.4

12.1 BRIEF OVERVIEW OF THE OBJECTIVES OF ENVIRONMENTAL CHEMISTRY

Environmental chemistry studies various aspects of chemical species in the environment, such as their sources, transport, reactions, effects, and their fate (Manahan, 1991, 2001). We present here a brief overview of the environmental chemistry objectives (Manahan, 1991, 2001; Schwarzenbach et al., 2003; Connell, 2005; Girard, 2005; van Loon and Duffy, 2011; Baird and Cann, 2012). Among numerous objectives, we select those that are best aligned with the objectives of this textbook, namely, organic and green chemistry. For this reason, we do not address inorganic or radioactive pollutants.

Environmental chemistry studies chemical processes that involve naturally occurring chemicals, as well as chemical pollutants of anthropogenic origins (anthropogenic—resulting from human activities) (Baird and Cann, 2012). Sometimes, there is an overlap between the two sources: natural and anthropogenic. One such example includes halogenated hydrocarbons (such as chloromethane) that are found in the atmosphere. These compounds may come from anthropogenic sources, but also from natural sources, such as from the biological activity of the oceans (Connell, 2005). In principle, the anthropogenic sources can be better controlled than the natural ones. Our focus is on the anthropogenic chemical pollutants, because these provide a challenge to green chemistry.

Environmental chemistry also studies natural ways chemicals are removed from the environmental systems in which they reside. These ways include abiotic chemical processes, such as hydrolysis, oxidation, and photochemical degradation, but also degradation by microbial enzymatic reactions. Environmental chemistry is also concerned with the ways chemical pollutants affect living organisms and ecosystems, and the ways the negative impact of these pollutants can be neutralized and reversed by natural means that are available to the ecosystems. Physical and chemical properties of various environments on the Earth influence abiotic chemical reactions, and are thus also studied.

It was environmental chemistry with its analytical techniques that detected and monitored the anthropogenic chemicals in the environment. Further, environmental chemistry applied toxicology principles to the study of toxic effects of anthropogenic chemicals on life and environment. A new branch of toxicology, ecotoxicology, was developed (Connell et al., 1999). Persistence and fate of chemical pollutants in the environment was also studied (Lipnick and Muir, 2001; Lipnick et al., 2001). Biodegradation, bioaccumulation, and structure–biodegradability tendencies of chemical pollutants also became a subject of study by environmental chemistry (Lipnick et al., 2001; Sijm, 2001; Schwarzenbach et al., 2003). These new areas of study broadened the objectives of environmental chemistry from the initial ones, namely, detection and monitoring of chemicals in the environment.

Environmental chemists study chemical pollutants in the soil, water, and air. Common pollutants in the soil are various pesticides, agricultural chemicals, and industrial chemicals. The latter are also found in water, commonly in wells close to hazardous waste sites. Pharmaceuticals, fire retardants, plasticizers, and other pollutants are also commonly found in water. Chemical pollutants are also found in the air. Some of these pollutants exhibit a negative impact on the ozone layer. Examples include halogenated hydrocarbons, notably chlorofluorocarbons, and various air pollutants resulting from industrial activity and products, especially volatile organic compounds (VOCs), which are often solvents. Ground-level air pollution is the result of the chemical reaction of primary pollutants, such as VOCs and NO, which react with oxygen under sunlight to give more toxic secondary pollutants, such as O_3, HNO_3, and various organics (Manahan, 2001; Baird and Cann, 2012; Connell, 2005).

Table 12.1 shows selected categories and examples of environmental chemical pollutants. More comprehensive and detailed coverage of this topic is available in the literature (Manahan, 1991, 2001; Schwarzenbach et al., 2003; Connell, 2005; Girard, 2005; van Loon and Duffy, 2011; Baird and Cann, 2012).

TABLE 12.1
Selected Categories and Examples of Common Environmental Chemical Pollutants

Categories of Chemical Pollutants	Examples of Pollutants, with Selected Information on Their Sources, Uses, and Effects on Life and Environment
Petroleum hydrocarbons in general	From drilling, refining, and consumption
Halogenated hydrocarbons in general	Pesticides
Polychlorinated biphenyls	From capacitors and large transformers
Dioxins	From chemical processes that involve chlorine, from combustion processes
Polybrominated diphenyl ethers	Used as fire retardants
Chlorofluorohydrocarbons and related compounds (e.g., hydrochlorofluorocarbons [HCFCs] and hydrofluorocarbons [HFCs])	From air conditioning and electronics industry; have ozone depletion potential
Perfluorinated sulfonates and related compounds	Surfactants, multiple uses
Perfluorinated alkyl acids	Used for coating surfaces for nonstick cooking-ware
Synthetic polymers in general	Polystyrene, a nondegradable polymer
Plasticizers	Phthalate esters, etc.
Monomers used for polymer production in general	Polyvinyl chloride production
Bisphenol A	A monomer used in production of polycarbonate plastics; exhibits estrogenic properties
Soaps and detergents in general	Phosphate pollution in lakes
Alkylbenzene sulfonates and alkyl sulfates	Used as detergents
Pesticides and agricultural chemicals in general	
Organochlorine insecticides	DDT, Lindane
Organophosphate insecticides	Malathion, Chlorpyrifos
Carbamate insecticides	Carbaryl, Pirimicarb
Herbicides in general	
Herbicides that are bipyridylium compounds (contain two pyridine rings)	Diquat, Paraquat
Herbicides that are triazines (heterocyclic nitrogen compounds that contain three nitrogens in the ring)	Arazine, Metribuzin
Herbicides that are substituted amides	Alachlor, Propanil

(Continued)

TABLE 12.1 (*Continued*)
Selected Categories and Examples of Common Environmental
Chemical Pollutants

Categories of Chemical Pollutants	Examples of Pollutants, with Selected Information of Their Sources, Uses, and Effects on Life and Environment
Herbicides that are chlorophenoxy compounds, nitroanilines, substituted ureas, carbamates, thiocarbamates, and other categories	2,4-D, Trifluralin, R-mecoprop, etc.
Organometallic compounds	Catalysts such as organoaluminum compounds, organomercury compounds from incineration of medical and municipal waste
Pharmaceuticals	See Table 12.2

Sources: Manahan, S. E., *Environmental Chemistry*, 5th edn., Lewis Publishers, Chelsea, MI, 1991; Manahan, S. E., *Fundamentals of Environmental Chemistry*, 2nd edn., Lewis Publishers, CRC Press, Boca Raton, FL, 2001; Connell, D. W., *Basic Concepts of Environmental Chemistry*, 2nd edn., CRC Press, Taylor & Francis Group, Boca Raton, FL, 2005; Girard, J. E., *Principles of Environmental Chemistry*, Jones and Bartlett Publishers, Sudbury, MA, 2005; Schwarzenbach, R. P. et al.: *Environmental Organic Chemistry*, 2nd edn. 2003. Copyright Wiley-VCH Verlag GmbH & Co. KGaA.; Van Loon, G. W. and Duffy, S. J., *Environmental Chemistry, A Global Perspective*, 3rd edn., Oxford University Press, Oxford, 2011; Baird, C. and Cann, M., *Environmental Chemistry*, 5th edn., W. H. Freeman and Company, New York, 2012.

The objectives of environmental chemistry have evolved over time, leading to its current interdisciplinary nature. The central thread is analytical chemistry, which is used for measuring and monitoring chemicals in the environment. With the ever-increasing development of analytical instruments and their sensitivity, environmental chemists can now detect and monitor chemical pollutants that are present in the environment in extremely small amounts. Some such micropollutants are pharmaceuticals, which exhibit biological activities even at very small amounts. Study of pharmaceuticals in the environment is an emerging area of environmental chemistry. We shall cover this topic in Section 12.3.

12.2 RELATIONSHIP BETWEEN ENVIRONMENTAL AND GREEN CHEMISTRY

Historically, environmental problems caused by the chemical pollutants acted as an impetus for the birth of green chemistry and its action (see Sections 1.1.1 and 1.1.2). In the past, chemical pollutants that were generated by the chemical industry were released to the environment believing that they would not be harmful if buried in the ground or diluted in water or by the atmosphere. This turned out not to be the case, as many of these chemicals became toxic pollutants (see Section 1.1.1).

Environmental chemistry provided an understanding of the complex and often detrimental effects of chemical pollutants on life and the environment. This is where green chemistry found its role. When we design industrial, pharmaceutical, and other chemicals that can find their way into the environment, we should design them to be green, and thus as harmless to the life and environment as possible.

Still, environmental chemistry will continue to inform green chemistry about the greening success, because it is not possible at this time to have a perfect *a priori* green chemical design. Some chemicals may show unanticipated negative effects on life and the environment, and then may have to be replaced with greener alternatives.

Environmental chemistry thus both informs and challenges green chemistry. The latter must respond to the findings of environmental chemistry by improving the greenness of chemicals.

Green chemistry had a substantial impact on the analytical methods that are used in environmental chemistry. Some methods generate toxic waste, such as reagents and organic solvents that may be used for the extractions during sample preparation. Greening of such procedures and of the analytical chemistry in general is described in Chapter 11. Specifically in terms of environmental analysis, greening of many of its aspects has also been accomplished. Representative green analytical methods for determining organic pollutants in waters, wastewaters, soil samples, and sediments are reviewed by Santelli et al. (2012).

We thus must use a tandem approach of green and environmental chemistry to solve environmental problems that are caused by anthropogenic chemical pollutants. Green and environmental chemistry are both evolving and remain interlinked.

In Section 12.3, we examine the presence and effects of pharmaceuticals in the environment.

12.3   PHARMACEUTICALS AS EMERGING ENVIRONMENTAL POLLUTANTS

A large number of pharmaceutical are used by humans. Many are also in veterinary use. These pharmaceuticals find their way into the environment by different routes. These include excretion of the original drug that was not completely absorbed, excretion of drug metabolites, and disposing of unused medications to trash, sewers, or septic tanks. Some pharmaceutical pollutants escape degradation in water treatment plants and then enter the environment. Pharmaceuticals may also come from pharmaceutical industry waste.

Table 12.2 shows selected categories and examples of common human drugs that are routinely detected in the aquatic environment (Khetan and Collins, 2007; Glassmeyer et al., 2008; Jjemba, 2008).

Most of the drugs have shorter half-lives in the environment than the traditional toxic pollutants, such as dichlorodiphenyltrichloroethane (DDT) or polychlorinated biphenyls (PCBs), which persist in the environment. However, pharmaceuticals are introduced continuously into the environment and are considered "pseudopersistent" (Ankley et al., 2007).

Pharmaceuticals in water, including drinking water, are a cause of special concern. Although their concentration may be below the targeted pharmacological effects, they are continuously replenished. This leads to a chronic exposure to low concentrations of biologically active compounds. Effects of such long-term exposure on health are mostly unknown. Such effects may result not only from the exposure to

TABLE 12.2

Selected Categories and Examples of Common Human Pharmaceuticals in Aquatic Environments

Drug Category	Specific Examples
Analgesics	Acetaminophen, ibuprofen, naproxen
Antibiotics	Ciprofloxin, erythromycin, sulfamethoxazole
Beta blockers (correct irregular heartbeat, relieve angina chest pain, reduce high blood pressure)	Propranolol, Metoprolol
Anti-hyperlipidemics (blood lipid-lowering agents)	Statins, e.g., atorvastatin, clofibrate
Antidepressants	Fluoxetine
Anti-epileptics	Carbamazepine
Steroid hormones	Contraceptive pill, 17α-ethinylestradiol, and other synthetic estrogens

Sources: Khetan, S. K. and Collins, T. J., Human pharmaceuticals in the aquatic environment: A challenge to green chemistry, *Chem. Rev.,* 107, 2319–2364, 2007; Glassmeyer, S. et al., Environmental presence and persistence of pharmaceuticals, in *Fate of Pharmaceuticals in the Environment and in Water Treatment Systems,* Aga, D. S., ed., CRC Press, Taylor & Francis Group, Boca Raton, FL, 2008, pp. 3–51; Jjemba, P. K.: *Pharma-Ecology: The Occurrence and Fate of Pharmaceuticals and Personal Care Products in the Environment.* 2008. Copyright Wiley-VCH Verlag GmbH & Co. KGaA.

an individual drug, but also from a combination effect with other drugs or chemical pollutants (Cleuvers, 2003; Ankley et al., 2007). For the drugs that exhibit the same or similar mode of action, the combination effect may be stronger than expected from any single drug in the mixture. Further, synergism cannot be excluded between various drugs, independent of their modes of action, may it be similar or dissimilar (Cleuvers, 2003).

Pharmaceuticals are typically developed for humans (or animals), which then represent their target organisms. However, pharmaceuticals may affect various nontarget organisms as well, often unexpectedly.

Some pharmaceuticals are extremely potent at very small doses. This is the case for synthetic estrogens, such as 17α-ethinylestradiol (EE2), which are typically used in contraceptive pills. EE2 as a pharmaceutical pollutant in water may affect aquatic organisms differently at different stages of their development, and may cause various developmental and reproductive problems (Ankley et al., 2007). For example, EE2 activates estrogen receptor and causes feminization of fish. Thus, it acts as a reproductive toxin. It targets a specific biological pathway, which is evolutionarily conserved across various species. It is an example of a pharmaceutical pollutant that has a relatively low acute toxicity, but exhibits a large chronic toxicity, which results in its reproductive toxicity. In such cases, the acute toxicity test alone may not be adequate for predicting ecotoxicity (Ankley et al., 2007). This example is not unique, as it applies also to other pharmaceuticals that act on the biological pathways that are preserved, and are thus not unique to humans. One such example is beta blocker propranolol.

Pharmaceuticals in the body are subject to metabolism. This produces extra compounds, so-called metabolites. These also constitute environmental pollutants (Arnold and McNeill, 2007). As one example, carbamazepine, an anti-epileptic drug, yields over 30 metabolites (Peréz and Barceló, 2008). Examples of metabolic pathways by which drugs are modified in the body include hydrolysis, for example, of esters such as aspirin (acetyl salicylic acid), and oxidation with cytochrome P450 enzymes. The latter are involved in metabolism of over half of all drugs (Arnold and McNeill, 2007).

Pharmaceuticals in the environment may be degraded by different pathways, abiotic or biotic. One of the abiotic pathways is photolysis, namely, degradation by sunlight. Some such decomposition products may be toxic, even more so than the original compounds. For this reason, photoproducts need to be identified and tested for toxicity. Other abiotic transformations include oxidation, for example, mediated by manganese oxide minerals, or coupling with organic matter in the environment, such as humic acid (Arnold and McNeill, 2007).

A significant way of degrading pharmaceuticals in the environment is biotic, which involves biotransformation, notably by various microbes (Schwarzenbach et al., 2003). In the living systems, chemical reactions are catalyzed by enzymes, which are tailored for specific substrates. However, substrate specificity is not perfect, and enzymes are able to act on compounds that are structurally similar to the substrate. Further, some enzymes are relatively nonspecific. They are thus able to target unwanted compounds, and chemically degrade them or derivatize them to a form that is suitable for excretion. Some such enzymes are not always present but are made on demand (are "induced") in response to the exposure to the harmful chemicals.

Degradation of pharmaceuticals also occurs in water treatment systems, in which chlorine, chlorine dioxide, or ozone are frequently used. They cause oxidative transformations and, in some pharmaceuticals, chlorination, for the former two agents. The treatment does not degrade all pharmaceuticals, and it gives toxic products in some cases (Khetan and Collins, 2007).

Presence of pharmaceuticals in the environment is expected to continue. This is an emerging problem that needs to be continuously monitored and addressed. Innovative solutions, such as replacing some drugs with the ecofriendly ones, would alleviate the problem.

12.4   DEVELOPMENT OF ECOFRIENDLY CHEMICALS IN RESPONSE TO THE FINDINGS OF ENVIRONMENTAL CHEMISTRY

Design of ecofriendly chemicals needs to be informed by the findings of environmental chemistry. In this section, we show selected ways for such a design.

Ecofriendly chemicals need to be degradable. This is stated in principle 10 of green chemistry: "Chemical products should be designed so that at the end of their function they do not persist in the environment and break down into innocuous degradation products" (Anastas and Warner, 1998). We show here the results of structure–biodegradability studies, which are directly applicable to the design of biodegradable chemicals.

Based on the analysis of an extensive empirical database on biodegradability of organic compounds, structure–biodegradability tendencies were established (Schwarzenbach et al., 2003). Chemicals were ranked on a biodegradability scale from "labile" to "recalcitrant." Biodegradability tendencies correlate well with the presence of specific functional groups and structural feature. Thus, biodegradable compounds typically have hydrolysable functional groups (carboxylic acid esters, amides, anhydrides, or phosphoric esters) or other specific groups (hydroxyl, formyl, and carboxy). Compounds that are not readily biodegradable contain chloro and nitro groups, particularly on aromatic rings, and have structural features such as quaternary carbons or tertiary nitrogens. However, these biodegradability tendencies are complicated by the fact that specific functional groups may influence biodegradability differently if oxygen is present (oxic conditions) or absent (anoxic conditions). Still, these relationships provide guidance for green chemical design.

Ecofriendly chemicals should not bioaccumulate. History has shown disastrous effects of many chemicals that bioaccumulate, such as DDT and related organochlorine insecticides. They bioaccumulate in many nontarget organisms, in which they cause toxic effects. Bioaccumulation is increased via food chain biomagnification.

Chemicals that bioaccumulate are typically persistent against biotransformations along the food chain. They are excreted at a very slow rate. In general, they exhibit low volatility and water solubility, high lipid solubility, and slow rate of both abiotic and biotic degradation. When designing new chemicals, one should consider such properties and learn from the negative examples of persistent organic chemical pollutants that bioaccumulate, such as DDT (Lipnick and Muir, 2001; Sijm, 2001).

A successful ecofriendly design of chemicals should start with the understanding of "retrospective ecotoxicology," namely, of the previously established data about toxic responses of chemical pollutants in the environment. The next step is to develop "predictive ecotoxicology," to enable prediction of the toxic effects of chemicals as we design them. The goal would be to design chemicals as benign as possible to life and environment so that they can be released safely into the environment.

Innovative solutions are needed in designing ecofriendly chemicals (Williams and Williams, 2012b). However, this is an extremely complex problem, with no simple solution. Systems thinking is needed both to grasp the problem and to find solutions.

REVIEW QUESTIONS

12.1 What are the two sources of chemicals in the environment?
12.2 Explain how environmental chemistry informs and challenges green chemistry.
12.3 What are "pseudopersistent" pollutants?
12.4 Are the acute toxicity tests adequate for prediction of ecotoxicity of pharmaceuticals?
12.5 Name a common enzyme that is involved in the oxidation of pharmaceuticals.
12.6 List two abiotic ways by which pharmaceuticals are degraded in the environment.
12.7 List five functional groups that enhance biodegradability.
12.8 List two structural features that diminish biodegradability.
12.9 List two typical requirements for chemicals to bioaccumulate.
12.10 List two typical physico-chemical features of chemicals that bioaccumulate.

ANSWERS TO REVIEW QUESTIONS

12.1 Natural and anthropogenic.

12.2 Environmental chemistry detects and monitors anthropogenic chemicals in the environment and discovers which ones are toxic to the ecosystems. This informs green chemistry that problem exists with such chemicals and poses a challenge that these chemicals need to be redesigned to become ecofriendly.

12.3 They degrade rather rapidly, and thus are not persistent, but are continuously introduced and are thus constantly present.

12.4 Not always. Some pharmaceutical pollutants, such as EE2, have a low acute toxicity but a large chronic toxicity.

12.5 Cytochrome P450.

12.6 Photolysis and oxidation by manganese oxide minerals.

12.7 –COOR, –CONR$_2$, –OH, –CHO, –COOH

12.8 Quaternary carbon and tertiary nitrogen.

12.9 They are persistent against biotransformations along the food chain and are excreted at a very slow rate.

12.10 Low water solubility, high lipid solubility.

REFERENCES

Aga, D. S., ed. (2008). *Fate of Pharmaceuticals in the Environment and in Water Treatment Systems*, CRC Press, Taylor & Francis Group, Boca Raton, FL.

Anastas, P. T. and Warner, J. C. (1998). *Green Chemistry: Theory and Practice*, Oxford University Press, Oxford.

Ankley, G. T., Brooks, B. W., Huggett, D. B., and Sumpter, J. P. (2007). Repeating history: Pharmaceuticals in the environment, *Environ. Sci. Technol.*, 41, 8211–8217.

Arnold, W. A. and McNeill, K. (2007). Transformation of pharmaceuticals in the environment: Photolysis and other abiotic processes, *Compr. Anal. Chem.*, 50, 361–385.

Baird, C. and Cann, M. (2012). *Environmental Chemistry*, 5th edn., W. H. Freeman and Company, New York.

Cleuvers, M. (2003). Aquatic ecotoxicity of pharmaceuticals including the assessment of combination effects, *Toxicolo. Lett.*, 142, 185–194.

Connell, D. W. (2005). *Basic Concepts of Environmental Chemistry*, 2nd edn., CRC Press, Taylor & Francis Group, Boca Raton, FL.

Connell, D. W., Lam, P., Richardson, B., and Wu, R. (1999). *Introduction to Ecotoxicology*, Blackwell Science, Oxford.

Cooper, E. R., Siewicki, T. C., and Phillips, K. (2008). Preliminary risk assessment database and risk ranking of pharmaceuticals in the environment, *Sci. Total Environ.*, 398, 26–33.

Daughton, C. G. (2007). Pharmaceuticals in the environment: Sources and their management, *Compr. Anal. Chem.*, 50, 1–58.

Girard, J. E. (2005). *Principles of Environmental Chemistry*, Jones and Bartlett Publishers, Sudbury, MA.

Glassmeyer, S., Kolpin, D.W., Furlong, E. T., and Focazio, M. J. (2008). Environmental presence and persistence of pharmaceuticals, in *Fate of Pharmaceuticals in the Environment and in Water Treatment Systems*, Aga, D. S., ed., CRC Press, Taylor & Francis Group, Boca Raton, FL, pp. 3–51.

http://epa.gov/ppcp/basic2.html, accessed on September 9, 2015.

https://en.wikipedia.org.wiki/Environmental_impact_of_pharmaceuticals_and_personal_care_products, assessed on September 9, 2015.

https://en.wikipedia.org/wiki/Environmental_chemistry, accessed on September 6, 2015.

https://en.wikipedia.org/wiki/Environmental_science, accessed on September 6, 2015.

Jjemba, P. K. (2008). *Pharma-Ecology: The Occurrence and Fate of Pharmaceuticals and Personal Care Products in the Environment*, John Wiley & Sons, Hoboken, NJ.

Khetan, S. K. and Collins, T. J. (2007). Human pharmaceuticals in the aquatic environment: A challenge to green chemistry, *Chem. Rev.*, 107, 2319–2364.

Kümmerer, K., ed. (2004). *Pharmaceuticals in the Environment, Sources, Fate, Effects and Risks*, 2nd edn., Springer-Verlag, Berlin, Germany.

Lipnick, R. L. and Muir, D. C. G. (2001). History of persistent, bioaccumulative, and toxic chemicals, in *Persistent, Bioaccumulative, and Toxic Chemicals I, Fate and Exposure*, Lipnick, R. L., Hermens, J. L. M., Jones, K. C., and Muir, D. C. G., eds., American Chemical Society Publishing, Washington, DC, pp. 1–12.

Lipnick, R. L., Hermens, J. L. M., Jones, K. C., and Muir, D. C. G., eds. (2001). *Persistent, Bioaccumulative, and Toxic Chemicals I, Fate and Exposure*, American Chemical Society Publishing, Washington, DC.

Manahan, S. E. (1991). *Environmental Chemistry*, 5th edn., Lewis Publishers, Chelsea, MI, p. 2.

Manahan, S. E. (2001). *Fundamentals of Environmental Chemistry*, 2nd edn., CRC Press, Boca Raton, FL.

Peréz, S. and Barceló, D. (2008). Advances in the analysis of pharmaceuticals in the aquatic environment, in *Fate of Pharmaceuticals in the Environment and in Water Treatment Systems*, Aga, D. S., ed., CRC Press, Taylor & Francis Group, Boca Raton, FL, pp. 53–80.

Santelli, R. E., Bezerra, M. A., Afonso, J. C., de Carvalho, M. F. B., Oliveira, E. P., and Freire, A. S. (2012). Environmental analysis, in *Handbook of Green Analytical Chemistry*, de la Guardia, M. and Garrigues, S., eds., John Wiley & Sons, Chichester, pp. 475–503.

Schwarzenbach, R. P., Gschwend, P. M., and Imboden, D. M. (2003). *Environmental Organic Chemistry*, 2nd edn., John Wiley & Sons, Hoboken, NJ.

Sijm, D. T. H. M. (2001). The "B" in PBT: Bioaccumulation, in *Persistent, Bioaccumulative, and Toxic Chemicals I, Fate and Exposure*, Lipnick, R. L., Hermens, J. L. M., Jones, K. C., and Muir, D. C. G., eds., American Chemical Society Publishing, Washington, DC, pp. 13–16.

Taylor, D. and Coombe, V. T. (2010). Environmental and regulatory aspects, in *Green Chemistry in the Pharmaceutical Industry*, Dunn, P. J., Wells, A. S., and Williams, M. T., eds., Wiley-VCH, Weinheim, Germany, pp. 83–100.

Van Loon, G. W. and Duffy, S. J. (2011). *Environmental Chemistry, A Global Perspective*, 3rd edn., Oxford University Press, Oxford.

Weinberg, H. S., Pereira, V. J., and Ye, Z. (2008). Drugs in drinking water, treatment options, in *Fate of Pharmaceuticals in the Environment and in Water Treatment Systems*, Aga, D. S., ed., CRC Press, Taylor & Francis Group, Boca Raton, FL, pp. 217–228.

Williams, R. T. and Williams, T. R. (2012a). Environmental science: Guiding green chemistry, manufacturing, and product innovations, in *Green Techniques for Organic Synthesis and Medicinal Chemistry*, Zhang, W. and Cue Jr., B. W., eds., John Wiley & Sons, Chichester, p. 33.

Williams, R. T. and Williams, T. R. (2012b). Environmental science; Guiding green chemistry, manufacturing, and product innovations, in *Green Techniques for Organic Synthesis and Medicinal Chemistry*, Zhang, W. and Cue Jr., B. W., eds., John Wiley & Sons, Chichester, pp. 33–66.

Zoller, U. (2005). Education in environmental chemistry: Setting the agenda and recommending action, *J. Chem. Ed.*, 82, 1237.

13 Greening Undergraduate Organic Laboratory Experiments

We believe that as the menu of green experiments grows in size and variety, the energy of activation for going green at more colleges and universities will be lowered, to the benefit of us all.

Thomas E. Goodwin (2009)

LEARNING OBJECTIVES

The learning objectives for this chapter are as follows.

Learning Objectives	Section Numbers
Learn about the literature sources on green organic laboratory experiments	13.1
Learn about methods of greening organic laboratory experiments	13.2

13.1 BRIEF SURVEY OF THE LITERATURE ON GREEN ORGANIC LABORATORY EXPERIMENTS

In this section, we describe selected literature sources on green organic laboratory experiments. The latter span virtually all types of experiments that are traditionally included in the undergraduate laboratory textbooks, but now they are greened. A large number of green experiments that are described in the literature facilitate the application of green chemistry in teaching modern undergraduate organic laboratories. Chemical equations for some of these experiments are shown in Figures 13.1 through 13.10.

Kirchhoff and Ryan (2002) edited the book *Greener Approaches to Undergraduate Chemistry Experiments*, which provides 14 innovative green experiments that are contributed by different authors. These experiments cover chemical concepts of electrophilic aromatic substitution, alkyne chemistry, organometallic chemistry, concerted reactions, enzyme catalysis, C–C bond formation, coordination chemistry, polymers, and other topics. Examples of specific experiments are electrophilic aromatic iodination of 4′-hydroxyacetophenone (Gilbertson et al., 2002) and benzoin condensation using thiamine as a catalyst instead of cyanide (Warner, 2002a) (see Figure 13.1).

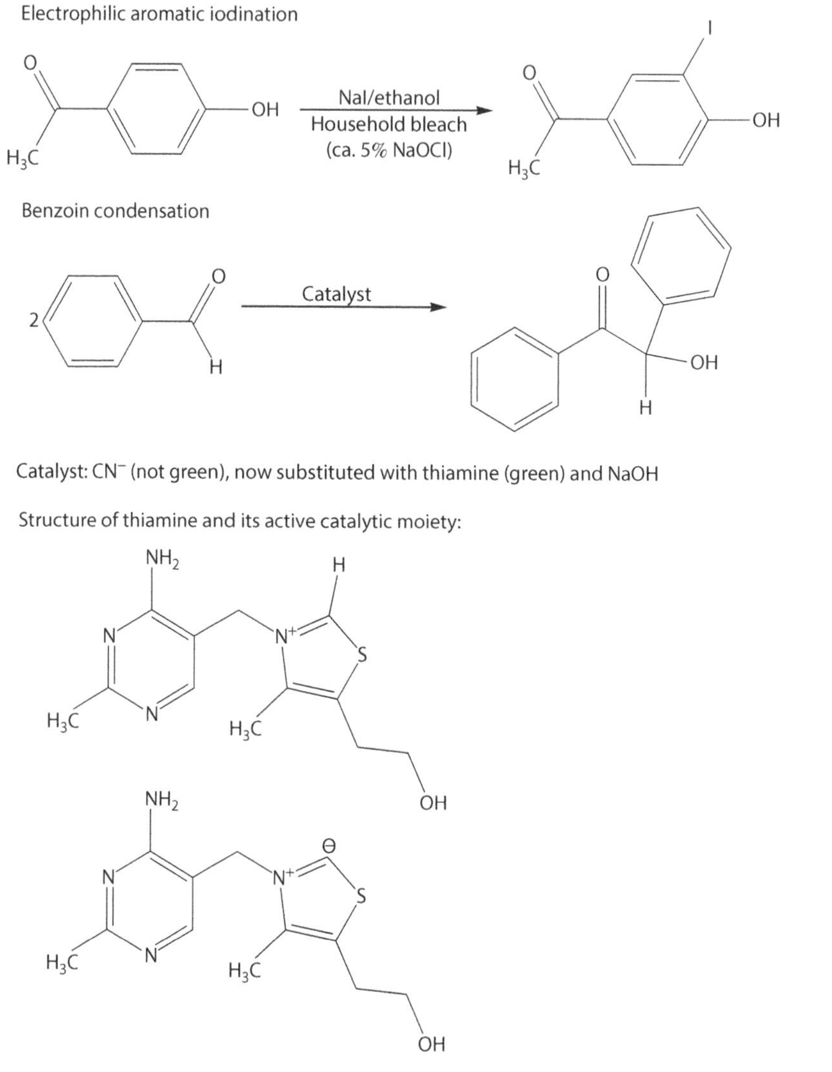

Electrophilic aromatic iodination

Benzoin condensation

Catalyst: CN⁻ (not green), now substituted with thiamine (green) and NaOH

Structure of thiamine and its active catalytic moiety:

FIGURE 13.1   Chemical equations for the green aromatic substitution and green benzoin condensation. (Kirchhoff, M. and Ryan, M.A., eds., *Greener Approaches to Undergraduate Chemistry Experiments*, Washington, DC. Copyright 2002 American Chemical Society; experiments by Gilbertson, R. et al., Electrophilic aromatic iodination of 4′-hydroxyacetophenone, in *Greener Approaches to Undergraduate Chemistry Experiments*, Kirchhoff, M. and Ryan, M. A., eds., Washington, DC, pp. 1–3. Copyright 2002 American Chemical Society; Warner, J. Benzoin condensation using thiamine as a catalyst instead of cyanide, in *Greener Approaches to Undergraduate Chemistry Experiments*, Kirchhoff, M. and Ryan, M. A., eds., Washington, DC, pp. 14–17. Copyright 2002a American Chemical Society.)

Doxsee and Hutchinson (2004) described 19 experiments in their book *Green Organic Chemistry, Strategies, Tools, and Laboratory Experiments*. These experiments include aldol condensation, benzoin condensation of furfural with thiamine as a

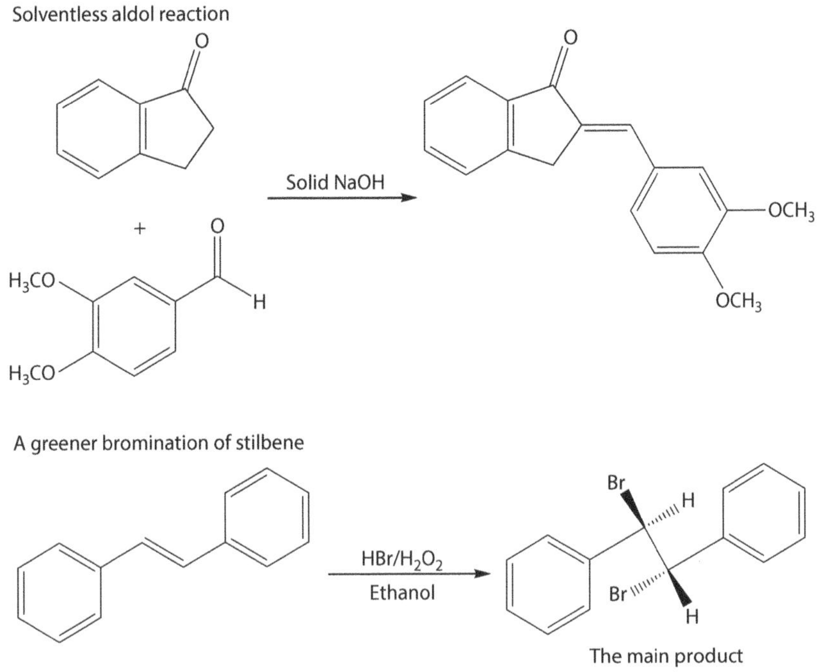

Solventless aldol reaction

A greener bromination of stilbene

The main product

FIGURE 13.2 Chemical equations for solventless aldol reaction and a greener bromination of stilbene. (From Doxsee, K. M. and Hutchinson, J. E., *Green Organic Chemistry, Strategies, Tools, and Laboratory Experiments*, Thomson Brooks/Cole, Belmont, CA, 2004.)

catalyst, bromination of stilbene, electrophilic iodination of vanillin, microwave synthesis of 5,10,15,20-tetraphenylporphyrin, resin-based oxidation, solid-phase photochemistry, and combinatorial chemistry. Examples of specific experiments are solventless reactions: the aldol reaction and a greener bromination of stilbene (see Figure 13.2).

Roesky and Kennepohl (2009) edited the book *Experiments in Green and Sustainable Chemistry*, which provides 46 experiments contributed by different authors. These are grouped into five topics, based on the green approach: (1) catalysis, (2) solvents, (3) high yield and one-pot syntheses, (4) limiting waste and exposure, and (5) special topics. We list here one experiment from each of the categories 1–5: chemoselective synthesis of acylals, ionic liquids as benign solvents, domino reactions for synthesis of natural products, green synthesis of the therapeutic agent juglone, and photochemistry for mild metal-free arylation reactions. Chemical equation for the green chemoselective synthesis of acylals that is performed without solvent is shown in Figure 13.3 (DeVore et al., 2009).

Ahluwalia (2012) authored the book *Green Chemistry, Environmentally Benign Reactions*, which lists a large number of the familiar named reactions, aspects of which are now greened. For example, these reactions may now be performed under microwave heating, under sonication, in the aqueous state, in the solid state, under solventless conditions, in the ionic liquids, in supercritical water, in supercritical CO_2, under enzymatic catalysis, with photochemical induction, and so on. Each of the reaction

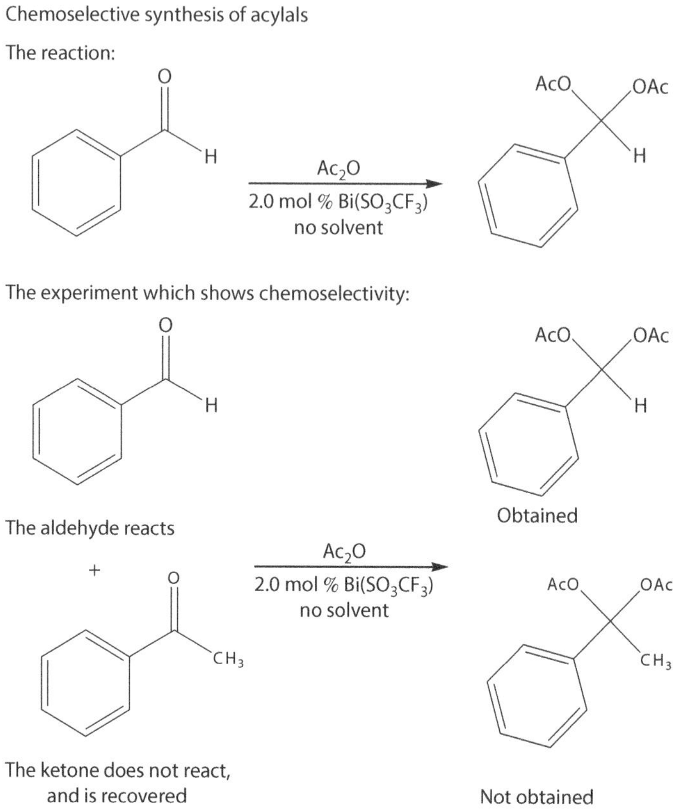

Chemoselective synthesis of acylals

The reaction:

The experiment which shows chemoselectivity:

The aldehyde reacts

Obtained

The ketone does not react, and is recovered

Not obtained

FIGURE 13.3   Chemical equation for green chemoselective synthesis of acylals. (DeVore, M. P. et al.: Chemoselective synthesis of acylals (1,1-diesters) from aldehydes: A discovery-oriented green organic chemistry laboratory experiment. *Experiments in Green and Sustainable Chemistry*. 2009. Roesky, H. W. and Kennepohl, D. K., eds. Copyright Wiley-VCH Verlag GmbH & Co. KGaA.)

descriptions specifies what is green about the reaction. Still, many aspects of these reactions may not be green. These require further greening. Out of many examples of the experiments from this book, we select the familiar Diels–Alder reaction, which has been described under the following green conditions: in water, in supercritical water, in supercritical CO_2, in ionic liquids, under microwave irradiation, and under sonication.

Leadbeater and McGowan (2013) published the book *Laboratory Experiments Using Microwave Heating*, which has 22 experiments, but not all of them are organic. We select Claisen rearrangement for preparation of allyl phenyl ether and Suzuki coupling reaction for preparation of biaryl. These reactions are green by virtue of using microwave heating. See Figure 13.4 for the equations for these reactions.

Organic laboratory textbooks are increasingly introducing green chemistry experiments. We give some examples.

Williamson and Masters (2011) included the following green experiments: solvent-less Cannizzaro reaction, Passerini reaction in water, benzoin condensation catalyzed

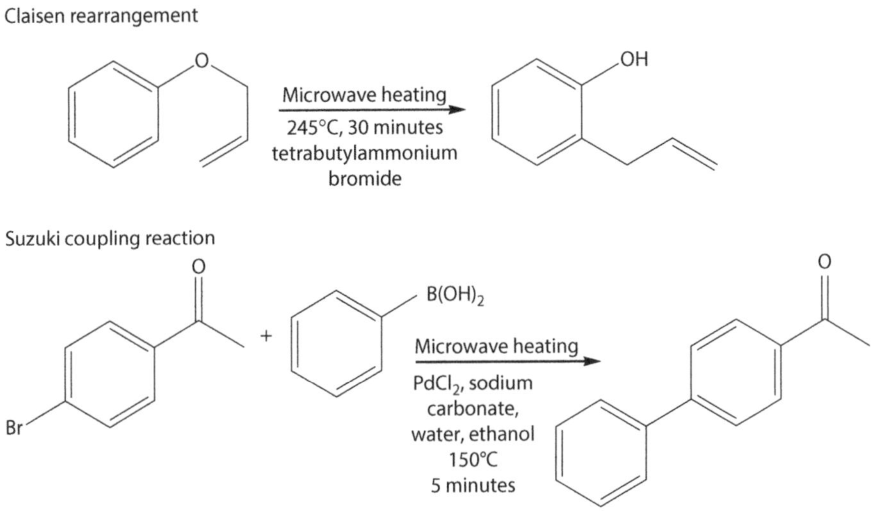

Claisen rearrangement

FIGURE 13.4 Chemical equations for the Claisen rearrangement and Suzuki coupling under microwave heating. (From Leadbeater, N. E. and McGowan, C. B., *Laboratory Experiments Using Microwave Heating*, CRC Press, Taylor & Francis Group, Boca Raton, FL, 2013. With permission.)

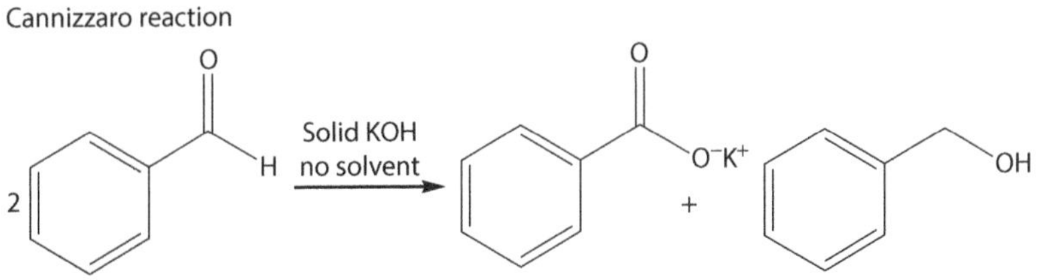

Cannizzaro reaction

FIGURE 13.5 Chemical equation for solventless Cannizzaro reaction. (From Williamson, K. L. and Masters, K. M., *Macroscale and Microscale Organic Experiments*, 6th edn., Thomson Brooks/Cole, Belmont, CA, 2011. With permission.)

with thiamine, hypochlorite oxidation of cyclohexanol to cyclohexanone, and air oxidation of fluorene to fluorenone. Figure 13.5 shows the equation for solventless Cannizzaro reaction. This experiment has a discovery element, because it challenges students to devise their own process for separation of products, benzoic acid and benzyl alcohol.

Palleros (2000) included Diels–Alder reaction in water and synthesis with baker's yeast as green experiments. Gilbert and Martin (2011) describe greener bromination of stilbene, shown in Figure 13.2, and emphasize the stereochemical aspect of this reaction. Numerous green experiments are available from various journals, notably *Journal of Chemical Education*, which also publishes supplementary material for each experiment. This material typically includes the infrared (IR) and nuclear magnetic resonance (NMR) spectra, melting point (mp's) or boiling point (bp's) of the products, average students' yields, and details about the experimental procedures.

In the References section, we give a selection of the journal articles that describe various green reactions. Examples include solventless reactions (Palleros, 2004; Cave and Raston, 2005; Phonchaiya et al., 2009), green oxidations (Straub, 1991; Hill et al., 2013; Leadbeater and Bobbit, 2014), Passerini reaction in water (Serafin and Priest, 2015), microwave synthesis of heterocyclic compounds (Katritzky et al., 2006), greening of the electrophilic aromatic substitution (Jones-Wilson and Burtch, 2005), biodiesel from plants (Pohl et al., 2012), and polymer experiments (Dintzner et al., 2012; Schneiderman et al., 2014).

These are just some of the examples of published green laboratory experiments. A comprehensive review of these is provided in the book *Green Organic Chemistry in Lecture and Laboratory*, edited by Dicks (2012). This book includes sources of experiments, such as textbooks, journals, and online sources. Numerous experiments are summarized, such as solventless reactions, reactions in water, reactions in green nonaqueous medium (e.g., in supercritical CO_2), oxidations, reductions, and halogenations with green reagents, and various green syntheses under microwave heating. This book has an extremely useful appendix, which lists numerous green reactions suitable for an undergraduate laboratory, together with the journal source, green aspects, and techniques that are involved. These reactions may not be perfectly green, however, and would thus benefit from further greening.

13.2 SELECTED METHODS OF GREENING ORGANIC LABORATORY EXPERIMENTS

In this section, we describe general methods for greening organic laboratory experiments and show their application on specific experiments.

Goodwin (2004) described an asymptotic approach to greening laboratory experiments. The term "asymptotic" describes well the greening process, because this process can never reach the target of the absolute greenness. Goodwin (2004) states that "all we can do is to strive to do better and make an asymptotic approach to the perfect green experiment."

In the examples that follow, we show some small steps that can improve the greenness of the specific experiments, some of which we have discussed in detail in Sections 4.2.1, 4.2.2, and 6.3. For example, we discussed the "on-water" Diels–Alder reaction (Section 4.2.1, Figure 4.2), as an example of a reaction that is green based on its perfect atom economy and the use of water as a solvent. We have also covered various factors that affect the reaction rate, such as the addition of prohydrophobic and antihydrophobic compounds (see Table 4.1). This particular reaction was adapted to an undergraduate experiment, with some minor structural variations (Palleros, 2000). It is performed as a green "on-water" reaction (Figure 13.6) (Palleros, 2000), but also under microwave heating (Figure 13.7) (Warner, 2002b; Leadbeater and McGowan, 2013).

Further greening of Diels–Alder reactions on water could be achieved by speeding it up by adding some of the prohydrophobic additives from Table 4.1. This type of greening is accessible to students, and they can become fully involved in such efforts within the lab period. For the purpose of greening, this experiment needs to be converted from a confirmatory one, in which the experiment confirms that the reaction

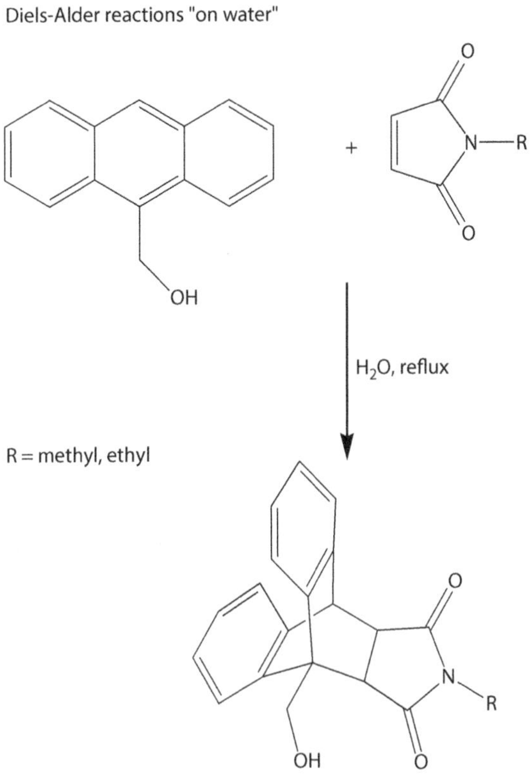

Diels-Alder reactions "on water"

R = methyl, ethyl

FIGURE 13.6 Diels-Alder on-water reactions that are often used as undergraduate organic experiments. (Palleros, D. R.: *Experimental Organic Chemistry*. 2000. Copyright Wiley-VCH Verlag GmbH & Co. KGaA; GEMS: Greener Education Material for Chemists, a website from the University of Oregon: http://greenerchem.uoregon.edu/gems.html. With permission.)

occurs as described in the literature, to an open-ended experiment, in which students continuously explore the influence of various additives on the reaction rate. Such an open-ended experiment can be done as a teamwork and be continued over a period of time, with a new group of students who will study different additives.

Importantly, involvement of students in greening the reactions will give them the ownership of the greening process and confidence that they too can do the greening. There is an additional benefit of the open-ended experiments. They are always fresh and interesting for both the students and the instructor. The results of the experiments are typically either not available in the literature or buried in various references, which students would have difficulty retrieving. Thus, the greening results would represent a discovery for students, which they always find exciting. For the instructor, the reality of teaching laboratory experiments is that the new experiments are costly to implement, and thus, the instructors end up teaching the same experiments over and over again, which becomes monotonous. However, the expansion from a confirmatory to an open-ended experiment enables the instructor to teach different aspects of the same experiment, and thus to keep teaching interesting.

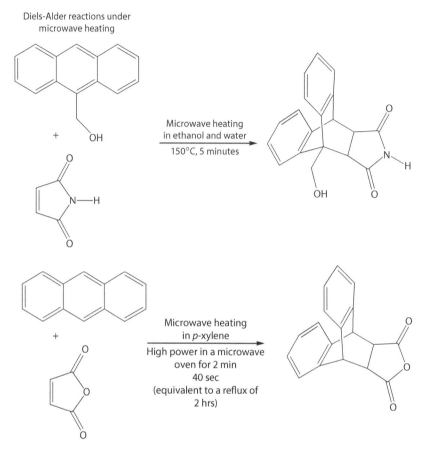

FIGURE 13.7 Examples of Diels–Alder reactions that are performed under microwave heating. (From Leadbeater, N. E. and McGowan, C. B., *Laboratory Experiments Using Microwave Heating*, CRC Press, Taylor & Francis Group, Boca Raton, FL, 2013; Warner, J. Microwave-assisted Diels-Alder reaction of anthracene and maleic anhydride, in *Greener Approaches to Undergraduate Chemistry Experiments*, Kirchhoff, M. and Ryan, M. A., eds., Washington, DC, pp. 8–10. Copyright 2002b American Chemical Society.)

We have also covered another on-water reaction, the Passerini reaction (Section 4.2.2, Figure 4.4). This reaction is also performed in the undergraduate laboratories, as an example of a green experiment, by virtue of a perfect atom economy and using water as a green solvent (e.g., Williamson and Masters, 2011; Serafin and Priest, 2015) (see Figure 13.8).

The original classical experiment (Figure 4.4 or Figure 13.8 for $R_1=R_2=H$) is adopted as an undergraduate experiment (e.g., Williamson and Masters, 2011). This experiment also provides an opportunity for expansion to an open-ended experiment. One question for the students could be to explore the scope of the reaction. After the confirmatory experiment is performed, students can explore reactivity of structurally modified starting materials to establish if they give the products that are

Passerini reaction

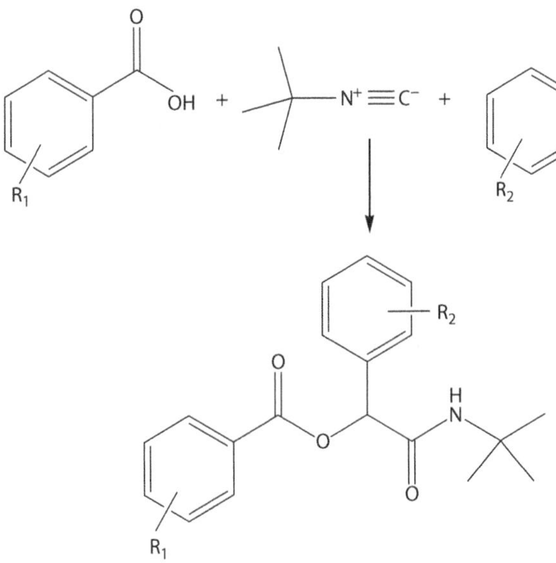

R_1, R_2: various combinations of –H, –OMe, –F, –Cl, Br, –NO$_2$, and –OH, in the *ortho, meta* or *para* positions;
Combinations that work the best (react quickly, form solid products that are easy to isolate, give spectra that are easy to interpret):
$R_1 = $ –H with $R_2 = $ –H, *ortho*–NO$_2$, *meta*–NO$_2$, *para*–NO$_2$, *ortho*–Cl, *para*–Br;
$R_1 = $ *meta*–OCH$_3$ with $R_2 = $ –H, *ortho*–NO$_2$, *meta*–NO$_2$.

FIGURE 13.8 Passerini reactions that are developed as undergraduate organic experiments. (From Williamson, K. L. and Masters, K. M., *Macroscale and Microscale Organic Experiments*, 6th edn., Thomson Brooks/Cole, Belmont, CA, 2011; Serafin, M. and Priest, O. P., Identifying Passerini products using a green, guided inquiry, collaborative approach combined with spectroscopic lab techniques, *J. Chem. Ed.*, 92, 579–581, 2015. With permission.)

analogous to that of the prototype reaction, and how such structural modifications affect reactivity. Serafin and Priest (2015) provided results of a series of Passerini reactions in which two starting materials, benzoic acid and benzaldehyde, are substituted with various electron-donating and electron-withdrawing groups in *ortho, meta*, and *para* positions (Figure 13.8). These authors emphasize also spectroscopic identifications of the products, and thus used as one of the criteria for the choice of substituents the ease of spectroscopic identification. However, their experiment can be used with many different variations and different emphases.

Thiamine-catalyzed benzoin condensation of benzaldehyde (Figure 13.1) is also adopted in many undergraduate laboratories (e.g., Gilbertson et al., 2002; Williamson and Masters, 2011). It works really well and is convincingly green, because the catalyst for the reaction has been greened dramatically. The original benzoin condensation used toxic cyanide ion as a catalyst, but it was found later that thiamine, which is vitamin B$_1$, does an equally good job as a catalyst. However, if one sets up an open-ended experiment and gives the students variously substituted benzaldehydes instead of benzaldehyde itself as a starting material, students will be surprised to find that only some of the substituted aldehydes condense, although there is no reaction with some other ones (e.g., with *p*-dimethylaminobenzaldehyde) (Ide and Buck, 1948; Smith and March, 2001). In the context of green chemistry, an important lesson is learned. Just because benzaldehyde reacts so well in this green reaction, the scope of the reaction itself may be limited. This limits its overall usefulness as a green reaction. In a subsequent guided inquiry study, students may be instructed to mix some of the substituted aldehydes that do not self-condense with benzaldehyde and try the reaction again. This time the reaction will work to give an unsymmetrical product containing moieties of both aldehydes. Actually, the benzoin condensation is more generally applicable to the preparation of unsymmetrical benzoins than to the preparation of symmetrical ones (Ide and Buck, 1948).

Another experiment that can be easily made into an open-ended experiment is the air oxidation of benzyl alcohols with a Cu(I)/2,2,6,6-tetramethylpiperidine-*N*-oxyl (TEMPO) catalyst system (Hill et al., 2013) (see Figure 13.9). This green oxidation experiment allows for an easy visualization of the reaction progress by color change, has an innovative catalytic cycle, and has an easy workup. However, the published experiment uses only the *para*-substituted benzyl alcohols. This experiment thus can

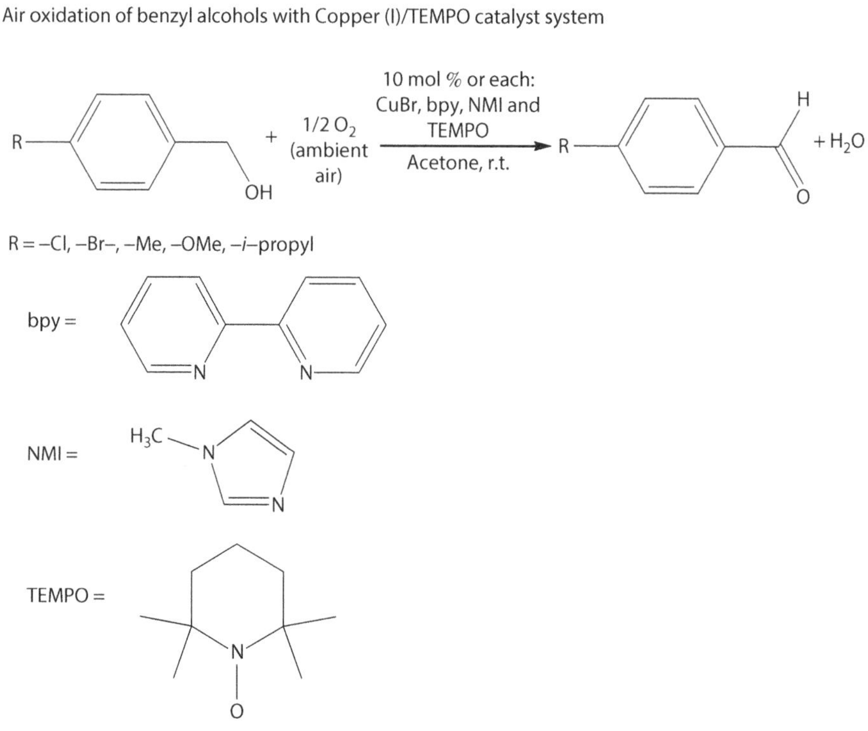

Air oxidation of benzyl alcohols with Copper (I)/TEMPO catalyst system

FIGURE 13.9 Chemical equation for the green oxidation of benzyl alcohols with a Cu (I)/ TEMPO catalyst system. (From Hill, N. J. et al., Aerobic alcohol oxidation using a copper (I)/TEMPO catalyst system, *J. Chem. Ed.*, 90, 102–105, 2013. With permission.)

also be converted to an open-ended experiment, in which students may be asked to use *ortho*-and *meta*-substituted benzyl alcohols to explore if these substrates react as fast as the *para*-ones and if they give comparable yields. The students may also explore the influence of substituents different than those reported in the original paper (Hill et al., 2013).

Another important point from the experiment by Hill et al. can be made to the students. In the published experiment, 1H-NMR is used as the primary spectroscopic tool for the structure identification. However, as we have learned in Table 11.1, the NMR is one of the least green instruments based on the energy consumption, whereas the IR is one of the greenest instruments. The students can be asked to perform structural identification based on the IR. This is rather easy, because distinct C=O band appears upon oxidation of benzyl alcohols to the corresponding aldehydes, whereas the –OH band from the alcohol disappears. The IR spectra of all common aldehydes that are the reaction products are available in the literature and could be used for spectra comparison. This type of greening may be warranted in some real-life applications, because the IR method is faster and cheaper, in addition to being greener.

A similar greening of the instrument use can be done in conjunction with the experiment on stereoselective synthesis of acylals, which is shown in Figure 13.3. The experiment shows that benzaldehyde forms acylal, whereas acetophenone does not. The experiment is set up such that a mixture of benzaldehyde and acetophenone is treated with the reagent for the acylal formation. The experiment calls for the use of NMR spectroscopy for identification of the material that is obtained after the workup. The NMR analysis confirms the presence of unreacted acetophenone and absence of benzaldehyde, which has reacted, and shows that the acylal product is formed. However, this analysis could be easily done by the IR. The starting benzaldehyde shows C=O frequency at 1703 cm^{-1}, acetophenone at 1686 cm^{-1}, and the acylal product at 1760 cm^{-1}. In the material that is obtained after the workup, one can easily see the C=O band for acetophenone, which did not react and is thus recovered, but the band for the C=O of benzaldehyde is not observed, because it has reacted. Separate experiment with only acetophenone shows that no reaction occurs (no product is obtained, and only recovered acetophenone is detected), whereas the reaction with benzaldehyde alone shows that the reaction does occur (only the product is obtained, and no unreacted benzaldehyde is observed). This simple IR analysis can be corroborated by the analysis of other characteristic IR bands (e.g., bands for the aldehyde H). Again, the IR method is faster, cheaper, and greener than the NMR one.

Chapter 6 provided examples of green solventless reactions (Figures 6.1 and 6.2). These and other solventless reactions are often performed as undergraduate experiments. For example, Palleros (2004) described solvent-free synthesis of a series of 20 different chalcones, out of which 17 can be obtained in minutes by mixing the starting materials, which comprise differently substituted benzaldehydes and acetophenones, with solid NaOH (see Figure 13.10).

One area of further greening of this experiment could be the recrystallization of the products. Solvents used for recrystallization were ethanol, ethanol–water, and toluene–ethanol. Although the yields of the crude products were high (81%–94%), the recovery of some of the products after recrystallization was low (10%–30%). Students could seek better solvents for recrystallization and try to eliminate toluene,

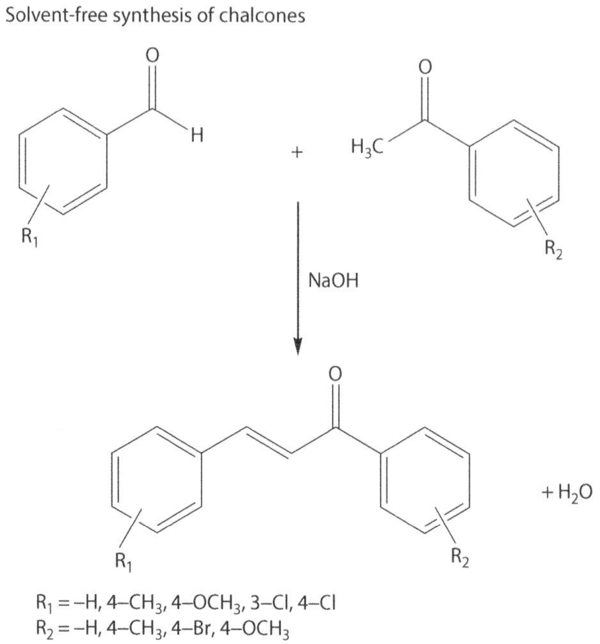

Solvent-free synthesis of chalcones

$R_1 = -H, 4-CH_3, 4-OCH_3, 3-Cl, 4-Cl$
$R_2 = -H, 4-CH_3, 4-Br, 4-OCH_3$

FIGURE 13.10 Chemical equation for solvent-free synthesis of chalcones. (From Palleros, D. R., Solvent-free synthesis of chalcones, *J. Chem. Ed.*, 81, 1345–1347, 2004. With permission.)

which is not a green solvent. This would make even the simple steps of recrystallization, such as making choice of solvent, a meaningful green chemistry exercise.

Chapter 7 showed a solid-state photochemical reaction of *trans*-cinnamic acid, which is also stereo selective (Figure 7.2). This reaction is another undergraduate green organic experiment (Doxsee and Hutchinson, 2004). Other substrates could be included in the photochemical study to explore the scope of the reaction.

The importance of the workup in greening the organic laboratory experiments needs to be emphasized (Dicks, 2015). Reaction workup needs to be evaluated for greenness as carefully as the reaction itself. The processes of isolation and purification of products can be greened by replacing toxic solvents that are used in recrystallization or chromatography with the green ones.

Proper training of the future chemists must include their early involvement in the efforts to green the laboratory experiments. Goodwin (2004) suggests that students include in their laboratory reports the answers to the following three questions:

1. What was green about the experiment?
2. What was not green?
3. How could the experiment be made greener?

REVIEW QUESTIONS

13.1 Class project: Explain the puzzling finding that benzoin condensation of some substituted benzaldehydes does not occur.

13.2 Class project: Examine some of your laboratory experiments in which you have performed recrystallizations. Which solvents did you use? Are they green? If not, can you find a greener substitute?

13.3 For the on-water Diels–Alder reaction (Figure 4.2, Section 4.2.1), it was stated that the hydrophobic effect forces a close proximity of the diene and dienophiles, which promotes the reaction. Could such proximity be achieved if we melt the two components and thus do not use water? Would the reaction occur in the melted state?

13.4 Consider the experiment that you are currently performing in the organic laboratory. Could it be made into an open-ended experiment? Which questions would you like to be answered by such an open-ended experiment?

ANSWERS TO REVIEW QUESTIONS

13.1 First research the mechanism of benzoin condensation, which is covered in most organic textbooks. Good sources are Smith and March (2001) and Williamson and Masters (2011). The mechanism is the same for the cyanide and thiamine catalysts. The basic mechanism is shown in Figure 13.11.

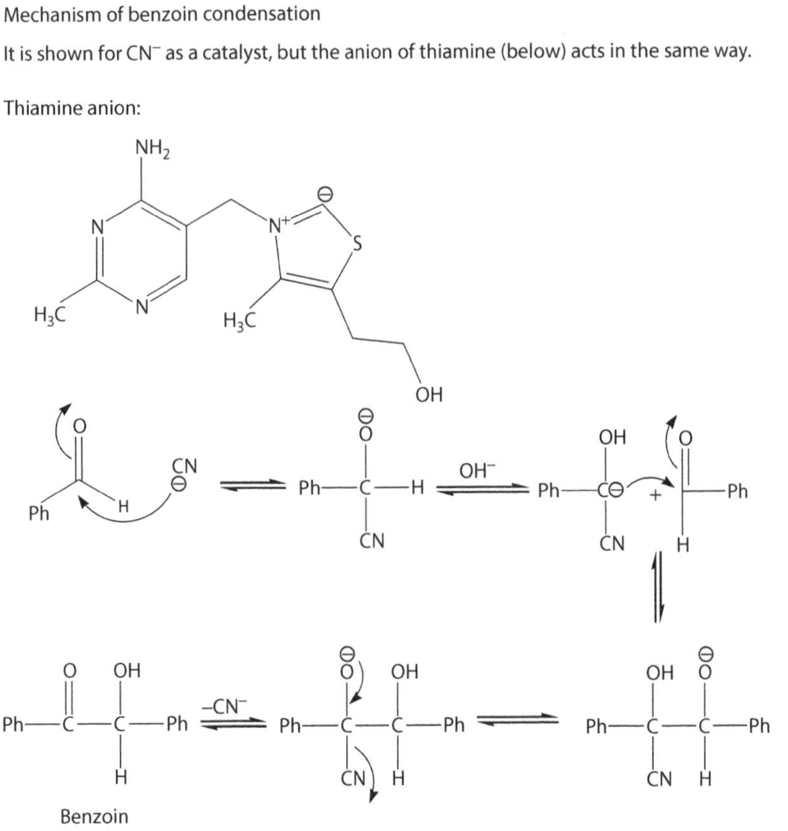

Mechanism of benzoin condensation

It is shown for CN⁻ as a catalyst, but the anion of thiamine (below) acts in the same way.

Thiamine anion:

FIGURE 13.11 A summary of the mechanism of benzoin condensation.

A list of differently substituted benzaldehydes that do not self-condense can be found in the work of Ide and Buck (1948). Consider the important mechanistic steps of the reaction, in which the substituent may be involved, such as influencing acidity of the aldehyde H after thiamine anion is added, and the nucleophilicity of the anion that will be formed after this H is removed by base. Note that *p*-nitrobenzaldehyde does not undergo benzoin condensation, because the nitro group reduces the nucleophilicity of the carbanion that is formed by the OH⁻. However, the strong electron-donating effect of *p*-dimethylamino group makes the loss of the proton to form carbanion very difficult. Thus, *p*-dimethylaminobenzaldehyde does not self-condense either (Williamson and Masters, 2011).

13.2 Answers will vary. However, students need to be reminded that the replacement solvent must not only be green but also be a good solvent for recrystallization. Thus, the compound to be recrystallized should dissolve in the hot solvent, but must not be soluble in the cold solvents. Students need to find solubility of the compound in the solvents under different temperatures. Merck Index is one possible source of such information. Mixed solvents are usually not as good choice as single solvents.

13.3 Yes. Yes.

13.4 The answers will vary. However, the instructor may assign a specific experiment for students to convert into an open-ended experiment.

REFERENCES

Ahluwalia, V. K. (2012). *Green Chemistry, Environmentally Benign Reactions*, CRC Press, Taylor & Francis Group, Boca Raton, FL.

Andraos, J. and Dicks, A. P. (2012). Green chemistry teaching in higher education: A review of effective practices, *Chem. Educ. Res. Pract.*, 13, 69–79.

Cann, M. C. and Dickneider, T. A. (2004). Infusing the chemistry curriculum with green chemistry using real-word examples, web modules, and atom economy in organic chemistry courses, *J. Chem. Ed.*, 81, 977–980.

Cave, G. W. V. and Raston, C. L. (2005). Green chemistry laboratory: Benign synthesis of 4.6-diphenyl[2,2′]bipyridine via sequential solventless Aldol and Michael addition reactions, *J. Chem. Ed.*, 82, 468–469.

DeVore, M. P., Huddle, M. G., and Mohan, R. S. (2009). Chemoselective synthesis of acylals (1,1-diesters) from aldehydes: A discovery-oriented green organic chemistry laboratory experiment, in *Experiments in Green and Sustainable Chemistry*, Roesky, H. W. and Kennepohl, D. K., eds., Wiley-VCH Verlag, Weinheim, Germany, pp. 50–52.

Dicks, A. P. (2015). Don't forget the workup, *J. Chem. Ed.*, 92, 405.

Dicks, A. P., ed. (2012). *Green Organic Chemistry in Lecture and Laboratory*, CRC Press, Taylor & Francis Group, Boca Raton, FL.

Dintzner, M. R., Kinzie, C. R., Pulkrabek, K., and Arena, A. F. (2012). The cyclohexanol cycle and synthesis of Nylon 6,6: Green chemistry in the undergraduate organic laboratory, *J. Chem. Ed.*, 89, 262–264.

Doxsee, K. M. and Hutchinson, J. E. (2004). *Green Organic Chemistry, Strategies, Tools, and Laboratory Experiments*, Thomson Brooks/Cole, Belmont, CA.

GEMS: Greener Education Material for Chemists, a website from the University of Oregon: http://greenerchem.uoregon.edu/gems.html, accessed on February 24, 2016.

Gilbert, J. C. and Martin, S. F. (2011). *Experimental Organic Chemistry, A Miniscale and Microscale Approach*, 5th edn., Thomson Brooks/Cole, Boston, MA. (See the following green experiments: Bromination of (E)-stilbene: The green approach, pp. 376–383; Oxidation of cyclododecanol to cyclododecanone, pp. 539–551.)

Gilbertson, R., Parent, K., McKenzie, L., and Hutchinson, J. (2002). Electrophilic aromatic iodination of 4′-hydroxyacetophenone, in *Greener Approaches to Undergraduate Chemistry Experiments*, Kirchhoff, M. and Ryan, M. A., eds., American Chemical Society, Washington, DC, pp. 1–3.

Goodwin, T. E. (2004). An asymptotic approach to the development of a green organic chemistry laboratory, *J. Chem. Ed.*, 81, 1187–1190.

Goodwin, T. E. (2009). The garden of green organic chemistry at Hendrix College, in *Green Chemistry Education, Changing the Course of Chemistry*, Anastas, P. T., Levy, I. J., and Parent, K. E., eds., American Chemical Society, Washington, DC, p. 37.

Hill, N. J., Hoover, J. M., and Stahl, S. S. (2013). Aerobic alcohol oxidation using a copper (I)/ TEMPO catalyst system, *J. Chem. Ed.*, 90, 102–105.

Ide, W. S. and Buck, J. S. (1948). The synthesis of benzoins, *Org. Reactions*, 4, 269–304.

Jones-Wilson, T. M. and Burtch, E. A. (2005). A green starting material for electrophilic aromatic substitution for the undergraduate organic laboratory, *J. Chem. Ed.*, 82, 616–617.

Kappe, C. D. (2004). Controlled microwave heating in modern organic synthesis, *Angew. Chem. Int. Ed.*, 43, 6250–6284.

Katritzky, A. R., Cai, C., Collins, M. D., Scriven, E. V., Singh, S. K., and Barnhardt, E. K. (2006). Incorporation of microwave synthesis into the undergraduate organic laboratory, *J. Chem. Ed.*, 83, 634–636.

Kirchhoff, M. and Ryan, M. A., eds. (2002). *Greener Approaches to Undergraduate Chemistry Experiments*, American Chemical Society, Washington, DC.

Leadbeater, N. E. and Bobbitt, J. M. (2014). TEMPO-derived oxoammonium salts as versatile oxidizing agents, *Aldrichimica Acta*, 47(3), 64–74.

Leadbeater, N. E. and McGowan, C. B. (2013). *Laboratory Experiments Using Microwave Heating*, CRC Press, Taylor & Francis Group, Boca Raton, FL.

Lehman, J. W. (2009). *Operational Organic Chemistry, A Problem-Solving Approach to the Laboratory Course*, 4th edn., Pearson, Prentice Hall, Upper Saddle River, NJ.

Palleros, D. R. (2000). *Experimental Organic Chemistry*, John Wiley & Sons, Hoboken, NJ. (See the following green experiments: Diels–Alder reaction in water, pp. 408–414; Asymmetric synthesis with Baker's yeast: An open-ended experiment, pp. 645–652).

Palleros, D. R. (2004). Solvent-free synthesis of chalcones, *J. Chem. Ed.*, 81, 1345–1347.

Phonchaiya, S., Panijpan, B., Rajviroongit, S., Wright, T., and Blanchfield, J. T. (2009). A facile solvent-free Cannizzaro reaction, *J. Chem. Ed.*, 86, 85–86.

Pohl, N. L. B., Streff, J. M., and Brokman, S. (2012). Evaluating sustainability: Soap versus biodiesel production from plant oils, *J. Chem. Ed.*, 89, 1053–1056.

Polshettiwar, V. and Varma, R. S. (2010). Non-conventional energy sources for green synthesis in water (microwave, ultrasound, and photo), in *Handbook of Green Chemistry*, Volume 5: Reactions in Water, Li, C.-J., ed., Wiley-VCH, Weinheim, Germany, pp. 273–290.

Roesky, H. W. and Kennepohl, D. K., eds. (2009). *Experiments in Green and Sustainable Chemistry*, Wiley-VCH Verlag, Weinheim, Germany.

Schneiderman, D. K., Gilmer, C., Wentzel, M. T., Martello, M. T., Kubo, T., and Wissinger, J. E. (2014). Sustainable polymers in the organic chemistry laboratory: Synthesis and characterization of a renewable polymer from δ-decalactone and L-lactide, *J. Chem. Ed.*, 91, 131–135.

Serafin, M. and Priest, O. P. (2015). Identifying Passerini products using a green, guided inquiry, collaborative approach combined with spectroscopic lab techniques, *J. Chem. Ed.*, 92, 579–581.

Smith, M. B. and March, J. (2001). *March's Advanced Organic Chemistry, Reactions, Mechanisms, and Structure*, John Wiley & Sons, New York, pp. 1243–1244.

Straub, T. S. (1991). A mild and convenient oxidation of alcohols, *J. Chem. Ed.,* 68, 1048–1049.

Varma, R. S. (February 1999). Solvent-free organic syntheses using supported reagents and microwave irradiation, *Green Chem.,* 43–55.

Warner, J. (2002a). Benzoin condensation using thiamine as a catalyst instead of cyanide, in *Greener Approaches to Undergraduate Chemistry Experiments*, Kirchhoff, M. and Ryan, M. A., eds., American Chemical Society, Washington, DC, pp. 14–17.

Warner, J. (2002b). Microwave-assisted Diels-Alder reaction of anthracene and maleic anhydride, in *Greener Approaches to Undergraduate Chemistry Experiments*, Kirchhoff, M. and Ryan, M. A., eds., American Chemical Society, Washington, DC, pp. 8–10.

Williamson, K. L. and Masters, K. M. (2011). *Macroscale and Microscale Organic Experiments*, 6th edn., Thomson Brooks/Cole, Belmont, CA. (See the following green experiments: The Cannizzaro reaction: Simultaneous synthesis of an alcohol and an acid in the absence of solvent, pp. 369–371; Multicomponent reactions: The aqueous Passerini reaction, pp. 699–701; The benzoin condensation: Catalysis by the cyanide ion and thiamine, pp. 655–660; Cyclohexanone from cyclohexanol by hypochlorite oxidation, pp. 356–368; Air-oxidation of fluorene to fluorenone, pp. 191–194).

Index

Note: Page numbers followed by 'f' and 't' denote figures and tables, respectively.